L'Organisation de Coopération et de Développement Économiques (OCDE), qui a été instituée par une Convention signée le 14 décembre 1960, à Paris, a pour objectif de promouvoir des politiques visant :

— à réaliser la plus forte expansion possible de l'économie et de l'emploi et une progression du niveau de vie dans les pays Membres, tout en maintenant la stabilité financière, et contribuer ainsi au développement de l'économie mondiale;
— à contribuer à une saine expansion économique dans les pays Membres, ainsi que non membres, en voie de développement économique;
— à contribuer à l'expansion du commerce mondial sur une base multilatérale et non discriminatoire, conformément aux obligations internationales.

Les Membres de l'OCDE sont : la République Fédérale d'Allemagne, l'Australie, l'Autriche, la Belgique, le Canada, le Danemark, l'Espagne, les États-Unis, la Finlande, la France, la Grèce, l'Irlande, l'Islande, l'Italie, le Japon, le Luxembourg, la Norvège, la Nouvelle-Zélande, les Pays-Bas, le Portugal, le Royaume-Uni, la Suède, la Suisse et la Turquie.

L'Agence de l'OCDE pour l'Énergie Nucléaire (AEN) a été créée le 20 avril 1972, en remplacement de l'Agence Européenne pour l'Énergie Nucléaire de l'OCDE (ENEA) lors de l'adhésion du Japon à titre de Membre de plein exercice.

L'AEN groupe désormais tous les pays Membres européens de l'OCDE ainsi que l'Australie, le Canada, les États-Unis et le Japon. La Commission des Communautés Européennes participe à ses travaux.

L'AEN a pour principaux objectifs de promouvoir, entre les gouvernements qui en sont Membres, la coopération dans le domaine de la sécurité et de la réglementation nucléaires, ainsi que l'évaluation de la contribution de l'énergie nucléaire au progrès économique.

Pour atteindre ces objectifs, l'AEN :

— *encourage l'harmonisation des politiques et pratiques réglementaires dans le domaine nucléaire, en ce qui concerne notamment la sûreté des installations nucléaires, la protection de l'homme contre les radiations ionisantes et la préservation de l'environnement, la gestion des déchets radioactifs, ainsi que la responsabilité civile et les assurances en matière nucléaire ;*
— *examine régulièrement les aspects économiques et techniques de la croissance de l'énergie nucléaire et du cycle du combustible nucléaire, et évalue la demande et les capacités disponibles pour les différentes phases du cycle du combustible nucléaire, ainsi que le rôle que l'énergie nucléaire jouera dans l'avenir pour satisfaire la demande énergétique totale ;*
— *développe les échanges d'informations scientifiques et techniques concernant l'énergie nucléaire, notamment par l'intermédiaire de services communs ;*
— *met sur pied des programmes internationaux de recherche et développement, ainsi que des activités organisées et gérées en commun par les pays de l'OCDE.*

Pour ces activités, ainsi que pour d'autres travaux connexes, l'AEN collabore étroitement avec l'Agence Internationale de l'Énergie Atomique de Vienne, avec laquelle elle a conclu un Accord de coopération, ainsi qu'avec d'autres organisations internationales opérant dans le domaine nucléaire.

FOREWORD

Argillaceous materials are characterized by favourable properties from the viewpoint of radioactive waste isolation. In particular their low permeability and high sorption capacity make them a very effective barrier. It is therefore logical that the utilisation of argillaceous materials for waste isolation is under consideration in a number of countries.

The most straightforward approach to the utilisation of argillaceous materials for waste isolation is emplacement of the waste in a shale or claystone formation on land. Another possibility is emplacement of the waste in argillaceous sediments under the ocean floor. It is also feasible to use clays or a mixture of clay and sand as an artificial barrier around waste containers placed in cavities excavated in a different geological formation. Finally, clays could be used in different ways to reduce the permeability of other formations or for backfilling, plugging and sealing of cavities, shafts, boreholes and fractures.

There are still some unknowns in relation to the use of argillaceous materials, particularly for the containment of heat generating wastes, since the effects of heat on clay minerals and on pore fluids under the confining pressures existing at depth are not fully understood.

Additional difficulties exist for the in situ measurement of permeability and radionuclides migration since meaningful experiments in such low-permeability materials require excessively long periods of time.

As part of its programme on geologic disposal of radioactive wastes, the NEA organised a Workshop aimed at bringing together specialists working on the various aspects of R & D on the use of argillaceous materials for radioactive waste isolation. These proceedings represent a record of the papers and discussions at this meeting.

AVANT-PROPOS

Les matériaux argileux sont caractérisés par des propriétés propices du point de vue de l'isolement des déchets radioactifs. En particulier, leur faible perméabilité et leur capacité de sorption élevée en font une barrière très efficace. Il est donc logique que l'utilisation des matériaux argileux pour l'isolement des déchets soit à l'étude dans un certain nombre de pays.

La façon la plus simple de concevoir l'utilisation des matériaux argileux pour l'isolement des déchets consiste à déposer les déchets dans une formation terrestre de schiste ou d'argile. Une autre possibilité serait de les déposer dans des sédiments argileux, sous le fond de la mer. Il est également possible d'utiliser des argiles ou un mélange de sable et d'argile comme barrière artificielle autour des conteneurs de déchets placés dans des excavations creusées dans une formation géologique différente. Enfin, on pourrait utiliser des argiles de diverses manières pour réduire la perméabilité d'autres formations ou pour remblayer, colmater et obturer des excavations, puits, forages et fissures.

Il existe encore certaines inconnues en ce qui concerne l'utilisation des matériaux argileux, en particulier pour le confinement de déchets calorigènes, car les effets de la chaleur sur les minéraux argileux et sur les fluides en milieu poreux en présence des pressions de confinement existant en profondeur, ne sont pas intégralement compris.

Il existe des difficultés supplémentaires dans le cas de l'exécution in situ de mesures de perméabilité et de migration des radionucléides, car l'exécution d'expériences significatives dans des matériaux aussi peu perméables exige des périodes de temps extrêmement longues.

Dans le cadre de son programme d'évacuation de déchets radioactifs dans des formations géologiques, l'AEN a organisé une réunion de travail dont l'objectif était de rassembler des spécialistes menant diverses activités de recherche et de développement relatives à l'utilisation des matériaux argileux pour l'isolement des déchets radioactifs. Le présent document constitue un compte rendu des communications présentées et des échanges de vues au cours de cette réunion.

TABLE OF CONTENTS

TABLE DES MATIÈRES

SESSION 3 - SEANCE 3

Chairman - Président : M. R.H. HEREMANS (Belgium)

SESSION 4 - SEANCE 4

Chairman - Président : Dr. N.A. CHAPMAN (United Kingdom)

SESSION 5 - SEANCE 5

Chairman - Président : Dr. T.F. LOMENICK (United States)

Session 1

Chairman-Président
Mr. L.R. DOLE
(United States)

Séance 1

SUMMARY OF REPLIES TO THE QUESTIONNAIRE ON R&D ACTIVITIES RELEVANT TO THE DISPOSAL OF RADIOACTIVE WASTES IN ARGILLACEOUS FORMATIONS

F. Gera
Division of Radiation Protection and Waste Management
OECD Nuclear Energy Agency

In February 1979 the NEA Secretariat distributed a questionnaire about ongoing R&D activities on clays relevant to radioactive waste disposal. During the subsequent months the following countries have replied to the questionnaire : Australia, Belgium, Canada, Italy, Norway, Sweden, Switzerland, United Kingdom and United States.

Some replies indicate no or moderate actuvities on clays ; for example : Australia, Norway and Sweden have no plans, at the moment, to investigate the possibility of waste disposal in argillaceous formation. On the other hand the Swedish disposal concept in crystalline rocks includes surrounding the waste canisters and back-filling the repository with a mixture of sand and bentonite. In this framework, a number of investigations have been carried out and are planned on the behaviour of sand-bentonite mixtures. The Swedish reply to the questionnaire consists in a list of references to these studies.

The most important activities described in the various questionnaires are summarised below.

Belgium

The feasibility of radioactive waste disposal in a clay formation of Oligocene age, called Boom clay, which underlies the Mol Center, is being investigated. Coring of the formation has been carried out and a number of tests and laboratory determinations have been performed on clay samples.

An in situ heating experiment at shallow depth is underway in a quarry south of Antwerp where the Boom clay outcrops.

A feasibility study of a waste repository capable of accommodating the waste generated by an installed nuclear capacity of 10.000 MWe during 30 years has been performed. Due to the great plasticity of the Boom clay, tunnels would require extensive support ; a cast iron lining has been considered.

Linear structures in northeastern Belgium have been identified on satellite photographs and investigated on the ground by means of high resolution seismic reflection surveys.

Canada

Some limited investigations of argillaceous sediments for the purpose of radioactive waste disposal have been recently started.

Sub-surface data on Upper Ordovician shale units in southern Ontario have been compiled. Laboratory tests and surface fracture mapping for this formation are being considered.

Clay minerals are being investigated for their potential as backfill and buffering materials. The main emphasis is on determination of the hydrothermal stability particularly in the 150 to 200°C temperature range.

Emplacement techniques are being considered as part of the repository conceptual design studies.

In addition, glacial till deposits in northern Ontario and possibly southern Ontario will be investigated in relation to their potential for disposal of low- and medium-level waste at shallow depth.

Italy

A marly-clay formation of Pliocene-Pleistocene age, located in southern Italy, has been investigated in order to assess its potential for radioactive waste disposal.

In situ heating experiments at shallow depths have been carried out to investigate the thermal response of the formation. The most recent experiment has given highly unexpected results since no direct relationship between temperature and distance from the heater was observed.

Heating experiments have also been performed in the laboratory in a large block of clay.

Finally, clay samples have been subjected to a number of chemico-physical determinations, including Kd's for Sr, Cs and Ce.

Switzerland

A regional literature study of three argillaceous formations has been carried out and areas deserving detailed field investigations outlined. These investigations will be complemented by a number of test galleries and boreholes (up to 2.000 m depth).

Construction of a field laboratory, and a number of in situ and laboratory experiments are in the planning stage.

The conceptual design and feasibility study for a waste repository in clay is underway.

United Kingdom

The Institute of Geological Sciences (IGS) has been asked to examine the feasibility of high-level waste disposal in argillaceous formations. A number of formations of Paleozoic or Triassic ages are being considered but no specific study site has yet been identified.

Younger clays of Jurassic, Cretaceous and Tertiary age are also considered for possible emplacement of low- and medium-level waste.

Some preliminary studies have been initiated on the use of bentonite mixtures as buffer or plugging material and on the sorption capacity of certain argillaceous sediments.

A high pressure/temperature laboratory is being set up with the aim of studying the hydrothermal behaviour of the formation during the high temperature stage of a repository. This laboratory will start investigating crystalline rocks but will later move on to argillaceous formations.

The analytical methods of predicting temperature fields, thermal stresses and induced water movement around a repository, presently under development, could be applied to argillaceous formations.

United States

The feasibility of radioactive waste disposal in argillaceous formations is being investigated in general terms throughout the US. This study is based on existing literature and data ; it should permit to identify specific areas or specific formations that deserve further study.

Some formations are being considered in greater detail ; they are :

- sedimentary rocks of the southeastern Triassic Basin ;
- Cretaceous or younger sediments of the southeastern Coastal Plains ;
- western Cretaceous shales.

An in situ heating experiment at shallow depth has been carried out in the Conasauga shales near Oak Ridge, Tennessee. Observed and predicted temperature values were in close agreement.

A study on mineralogical changes in argillaceous materials that result from exposure to elevated temperature and pressure in natural conditions has been carried out. The results of this work have been published in report ONWI-21.

In the framework of the Waste Isolation Safety Assessment Program (WISAP) the sorption capacity of many rock types for a number of radionuclides has been determined. Data from sorption studies are being compiled in a data bank at Battelle Pacific Northwest Laboratory.

UNITED STATES NATIONAL WASTE TERMINAL STORAGE ARGILLACEOUS ROCK STUDIES*

George D. Brunton
Oak Ridge National Laboratory
Oak Ridge, Tennessee 37830 USA

ABSTRACT

The past and present argillaceous rock studies for the U.S. National Waste Terminal Storage Program consist of: (1) evaluation of the geological characteristics of several widespread argillaceous formations in the United States; (2) laboratory studies of the physical and chemical properties of selected argillaceous rock samples; and (3) two full-scale in situ surface heater experiments that simulate the emplacement of heat-generating radioactive waste in argillaceous rock.

*Research sponsored by the U.S. Department of Energy, under contract W-7405-eng-26 with the Union Carbide Corporation and administered by the Office of Nuclear Waste Isolation, Battelle Memorial Institute.

1. INTRODUCTION

As a result of recommendations by the United States (U.S.) National Academy of Sciences-National Research Council Advisory Committee in 1957 [1], the first active studies for the geological disposal of radioactive waste in the U.S. emphasized bedded salt as a repository medium. In the early 1970s, after completion of the underground vault test in a bedded-salt mine in Kansas, the U.S. Atomic Energy Commission (USAEC) broadened its objectives to include preliminary studies of other rock units within the U.S. which might be acceptable for terminal storage of radioactive waste. This work has gradually expanded - with intervals of minimal funding support - to include geological investigations, laboratory studies, and in situ experiments in several nonevaporite types of rock.

2. GEOLOGICAL INVESTIGATIONS

In 1971, the USAEC funded the U.S. Geological Survey (USGS) to summarize the existing geological and hydrological information on selected geographic areas and rock types that might be suitable for further investigation as geological repository sites. Shale, mudstone, and claystone of marine origin in areas of minimal structural deformation were considered to be most favorable. These rocks include the Ohio Shale of Devonian age in northern Ohio, and the Devonian-Mississippian Ellsworth Shale and the Mississippian Coldwater Shales in Michigan. The Pierre Shale and other thick shales of Cretaceous age in the Rocky Mountains were also considered as potential host rocks [2].

In 1976, the USGS, with funding from the U.S. Energy Research and Development Administration (USERDA), established a set of criteria that was used to select three geographical areas - one each in North Dakota, South Dakota, and Colorado - which contain units of the Pierre Shale that might be suitable for detailed investigation [3]. The USGS is continuing these studies with direct funding from the U.S. Department of Energy (DOE).

J. Droste and C. Vitaliano [4,5] reviewed the geological literature of the Illinois Basin in Illinois and Indiana and recommended several geographic areas in South Central Indiana where the Devonian New Albany Shale and the Ordovician Maquoketa Shale meet the rudimentary stratigraphic criteria for a first level screening for potential repository sites.

A similar review [6] of the Black Warrior Basin delineates geographical areas of central Mississippi where the depths and thicknesses of argillaceous rock merit detailed investigation. These formations are Cretaceous and younger. They are similar in composition to the Pierre Shale.

The Triassic basins of the eastern coast of the U.S. have been under investigation by the DOE Savannah River Plant since 1971 [7]. This element of the Bedrock Waste Storage Project is now being funded as a National Waste Terminal Storage (NWTS) project. The objective of this project is to "conduct regional reconnaissance studies of igneous and metamorphic rocks of the Piedmont Province, Triassic mudstones, shale, and sandstones, and the unconsolidated-to-semiconsolidated sands and clays of the Coastal Plains in areas of the coastal states east of the Blue Ridge Mountains between Maryland and Georgia" [8].

The Office of Waste Isolation (OWI) and the Office of Nuclear Waste Isolation (ONWI) have also funded a review of the Triassic basins in the southeastern U.S. [9], and a regional characterization of the Devonian shales in the eastern U.S. [10] for the storage or disposal of radioactive wastes respectively.

3. LABORATORY STUDIES

Most of the laboratory studies of argillaceous rocks for waste disposal are limited-objective by-products of broader NWTS programs such as the Waste Isolation Safety Analysis Program (WISAP) of Battelle Pacific Northwest Laboratory (BPNL) and the waste/rock interaction studies at the Pennsylvania State University. These programs have long-range objectives that are not specific to particular rock types and will take years to complete.

Current available information on the thermal properties of clays and shales [11] and a classification of fine-grained sheet silicate rocks [12] have been published by OWI. The latter report is an in-depth discussion of the complexities of rocks that are commonly grouped together as shales.

The School of Geophysical Sciences, Georgia Institute of Technology, is establishing geothermometry techniques for determining the thermal history of argillaceous rocks, and the geochemical, mineralogical and structural changes that occur in shales subjected to postdepositional temperatures up to 400°C. The tools that are being developed for this purpose are mineralogical alteration assemblages, K/Ar ratios, $^{16}O/^{18}O$ ratios, organic metamorphism, and conodont color alteration. These tools are being applied to the Conasauga Shale in Tennessee and Georgia for field verification of the concept. Dr. C. E. Weaver has also completed a comprehensive review of the diagenesis of clay minerals in argillaceous rocks for ONWI [13].

Dr. I. W. Marine at the Savannah River Laboratory is investigating the effects of natural osmotic membranes on natural hydraulic gradients in argillaceous formations.

The Oak Ridge National Laboratory (ORNL) has recently started a project to evaluate three broad classes of argillaceous rocks for nuclear waste containment. These classes are defined as follows: (1) argillaceous rocks with no carbonaceous material and with little or no hydrous expanded clay mineral; (2) argillaceous rocks with carbonaceous material; and (3) argillaceous rocks with smectite as a major clay mineral constituent. Each class will be sampled over a range of values for the content of the major mineralogical constituents (i.e., clay, quartz and calcite).

Samples of the argillaceous rocks will need to be obtained from the Illinois or Appalachian basins and from Cretaceous or younger formations at sites where "typical" examples of these rocks can be obtained from well-characterized geological formations. Each sample will consist of ten 100-ft (33-m)-long cores of unweathered rock from a single site. Each sample will be divided into subsamples that will be characterized by: (1) complete chemical and mineralogical analyses, including measurement of the total volatile content of each sample over a range of temperatures from ambient to 900°C in 100°C steps; (2) response or reaction to large doses of gamma radiation; (3) measurement of the thermal properties of each sample at ambient temperature and 300°C; (4) measurement of the adsorptive properties of the rocks; (5) measurement of the mechanical properties at ambient temperature and 300°C; and (6) determination of the products of the interaction of the samples with simulated nuclear waste.

4. IN SITU EXPERIMENTS

Two in situ surface experiments have been conducted in argillaceous rocks. The first of these was designed and operated by Sandia Laboratories in the Conasauga Formation at the Oak Ridge Reservation in Tennessee [14,15]. The second experiment was also operated by Sandia at the Nevada Test Site in the Eleana Argillite [16,17]. The preliminary results of these experiments will be reported at the NEA/OECD Workshop on Argillaceous Materials for Isolation of Radioactive Waste.

5. REFERENCES

(1) National Academy of Sciences-National Research Council: Disposal of Radioactive Wastes on Land, Publ. 519, Washington, D.C. 1957.

(2) Merewether, E. A. et al.: Shale, Mudstone, and Claystone as Potential Host Rocks for Underground Emplacement of Waste, USGS-4339-5, U.S. Geological Survey Open-file Report, 1973.

(3) Shurr, G. W.: The Pierre Shale, Northern Great Plains; A Potential Isolation Medium for Radioactive Waste, USGS-77-776, U.S. Geological Survey Open-file Report, 1977.

(4) Droste, J. B., and C. J. Vitaliano: Geological Report of the Maquoketa Shale, New Albany Shale, and Borden Group Rocks in the Illinois Basin as Potential Solid Waste Repository Sites, Y/OWI/SUB-7062/1, Office of Waste Isolation, Union Carbide Corporation, Nuclear Division, Oak Ridge, Tennessee, June 1976.

(5) Droste, J. R.: Paleozoic Stratigraphy of Two Areas in Southeastern Indiana, Y/OWI/SUB-7062/2, Office of Waste Isolation, Union Carbide Coporation, Nuclear Division, Oak Ridge, Tennessee, September 1976.

(6) Mellon, F. F.: Basal Ottawa Limestone, Chattanooga Shale, Floyd Shale, Porters Creek Clay, and Yazoo Clay in Parts of Alabama, Mississippi, and Tennessee as Potential Host Rock for Underground Emplacement of Waste, Y/OWI/SUB-76/87950, Office of Waste Isolation, Union Carbide Corporation, Nuclear Division, Oak Ridge, Tennessee, February 1976.

(7) Parsons, Brinckerhoff, Quade and Douglas, Inc.: "United States Atomic Energy Commission Savannah River Plant Bedrock Waste Storage Project," Interim Preliminary Conceptual Analysis Report, Supplement No. 1, prepared for E. I. du Pont de Nemours & Co., Wilmington, Delaware, November 1972.

(8) Marine, I. W.: "Southeast Disposal Site Studies," Technical Progress Report for the Quarter October 1-December 31, 1978, ONWI-9(1), Office of Nuclear Waste Isolation, Battelle Memorial Institute, Columbus, Ohio, 1979.

(9) Weaver, C. E.: Waste Storage Potential of Triassic Basins in Southeast United States, Y/OWI/SUB-7009/2, Office of Waste Isolation, Union Carbide Corporation, Nuclear Division, Oak Ridge, Tennessee, July 1976.

(10) Lomenick, T. F., and R. B. Laughon: A Regional Characterization of the Devonian Shales in the Eastern U.S. for the Storage/Disposal of Radioactive Wastes," Proc. Workshop on the Use of Argillaceous Materials for the Isolation of Radioactive Waste, NEA/OECD, Paris 1979.

(11) Weaver, C. E.: Thermal Properties of Clays and Shales, Y/OWI/SUB-7009/1, Office of Waste Isolation, Union Carbide Corporation, Nuclear Division, Oak Ridge, Tennessee, July 1976.

(12) Weaver, C. E.: Fine-Grained Sheet Silicate Rocks, Y/OWI/SUB-7009/4, Office of Waste Isolation, Union Carbide Corporation, Nuclear Division, Oak Ridge, Tennessee, September 1977.

(13) Weaver, C.E.: Geothermal Alteration of Clay Minerals and Shales: Diagenesis, ONWI-21, Office of Nuclear Waste Isolation, Battelle Memorial Institute, Columbus, Ohio, August 1979.

(14) Krumhansl, J. L.: Preliminary Results Report Conasauga Near-Surface Heater Experiment, SAND79-1745, Sandia Laboratories, Albuquerque, New Mexico, June 1979.

(15) Krumhansl, J. L., and W. D. Sundberg: "The Conasauga Near-Surface Heater Experiment Implications for a Repository Sited in a Water Saturated Argillaceous Formation," Proc. Workshop on the Use of Argillaceous Materials for the Isolation of Radioactive Waste, NEA/OECD, Paris 1979.

(16) Lappin, A. R., and W. A. Olsson: "Material Properties of Eleana Argillite Extrapolated to Other Argillaceous Rocks, and Implications for Waste Management," Proc. Workshop on the Use of Argillaceous Materials for the Isolation of Radioactive Waste, NEA/OECD, Paris 1979.

(17) McVey, D. F., A. R. Lappin, and R. K. Thomas: "Test Results and Support Analyses of a Near-Surface Heat Experiment in the Eleana Argillite," Proc. Workshop on the Use of Argillaceous Materials for the Isolation of Radioactive Waste, NEA/OECD, Paris 1979.

Discussion

L.R. DOLE, United States

You have mentioned in your presentation that three sites for further study have been selected in the Pierre Shale by the United States Geological Survey. What were the criteria for selecting these sites ?

G.D. BRUNTON, United States

The criteria were :

- minimum depth - 300 m ;
- minimum thickness - 150 m ;
- low permeability ;
- easy accessibility ;
- low population density ;
- low or limited topographic relief ;
- government ownership of the land.

C.R. WILSON, United States

Are any studies being carried out in the US on highly plastic clays similar to the Boom clay in Belgium ?

G.D. BRUNTON, United States

Yes. The Pierre Shale is frequently bentonitic and semi-plastic and the Yazoo clay in Mississippi is a plastic clay.

F. GERA, NEA

The marine clays that are investigated in the framework of the Seabed Working Group are very plastic.

G.D. BRUNTON, United States

The investigations of marine clays are outside the scope of the National Waste Terminal Storage Program, which only addresses disposal in continental rocks. However, Mr. Krumhansl, of Sandia Laboratories, will report on the studies of marine clays.

A. BRONDI, Italy

You mentioned studies for determining the thermal history of argillaceous rocks and ensuing geochemical, mineralogical and structural changes in shales subjected to post depositional temperatures up to 400°. Were the high temperatures regional or were they related to localized features, such as a batholite, a laccolite or a geothermal field ?

G.D. BRUNTON, United States

The origin of the heat was regional. The high temperatures were related to depth of burial.

S. GONZALES, United States

I wist to comment about Dr. Brondi's question relative to the purpose of geothermometry studies at the Georgia Institute of Technology. The question dealt with determination of past high temperatures due to either metamorphism, or normal postdepositional processes.

Conodont color alteration and organic metamorphism, the latter based upon vitrinite-reflectance studies, are techniques derived by the petroleum industry to ascertain the thermal history of sediments, since that is an important factor in determining maturation of organic matter relative to hydrocarbon genesis. Thus, use of these techniques in studies carried out at Georgia Tech is for the purpose of studying the thermal history of any shale, and will most likely reveal paleotemperatures due only to normal postdepositional factors, i.e. deep burial and geothermal gradient.

A.A. BONNE, Belgium

In your paper you have mentioned that ORNL has classified clay formations in three classes. My question is the following : does that mean that ORNL people consider that each class of clay may be used for a specific type of waste ?

G.D. BRUNTON, United States

We do not think that all three classes of argillaceous rocks will be equally suitable for heat-generating waste. Our intention is to establish a data base for each class for the purpose of being able to decide which class or classes will be most suitable for each type of nuclear waste.

AN EXPERIMENT ON THE HEAT TRANSMISSION IN A CLAY ROCK

E. Tassoni
Comitato Nazionale per l'Energia Nucleare
CSN - Casaccia - S.M. di Galeria
Roma (Italy)

Abstract

The results of an heating experiment in clay, executed by means of an electrical heater simulating a canister containing high level radioactive waste, are reported. The experiment has been conducted in the Nuclear Centre of Trisaia (Southern Italy). The heater was placed at a depth of 25 meters and its power was 1.2 kW. The code used for the theoretical calculations has been the Heating 5, usually employed for thermal conduction studies. The temperature distribution at the surface of the heater resulted to fit well the theoretical values. On the other hand, the same fit did not occur for the temperature distribution in the clay formation. In the external zone of the experiment some measured temperatures have shown higher values than in proximity of the heater. Different causes have been examined to explain this anomalous behaviour of the clay and the most probable may be attributed to mechanical disturbance induced by drilling operation.

This work has been performed in the frame of the indirect programme "Management and Storage of Radioactive Waste" of the European Atomic Energy Community.

1. INTRODUCTION

The dissipation of heat arising from the radioactive decay constitutes an important problem of the geologic disposal of high level radioactive waste. The heat may affect the physical, chemical and mineralogical conditions of the host rock and deeply modify the natural context. CNEN (the Italian Atomic Energy Commission) has carried out an in situ heating experiment in a clay formation under the "Trisaia" Nuclear Research Centre, in Southern Italy. This experiment has been conducted in the framework of a contract between CNEN and CEC (Commission of European Communuties).

2. AIM OF THE EXPERIMENT

The experiment was directed to the following objectives :

a) to investigate the temperature increase in the clay, as induced by an electric heater simulating a canister containing high level radioactive wastes ;

b) to select materials and instrumentation for future similar experiments ;

c) to verify the fit between experimental data and theoretical calculations of the induced temperature increase in the clay.

3. POSITIONING AND DRILLING OF BOREHOLES

The geometrical parameters of the experiment have been defined using the Heating 5 code, already amployed in the study of the heat transmission in rock salt [1].

This code enables the theoretical distribution of the temperature around the canister to be determined.

A canister generating constant power has been assumed, with a surrounding homogeneous and isotropic clay medium, which has the following physical properties :

Thermal conductivity (cal/hr cm°C)	Specific gravity (g/cm^3)	Specific heat (cal/g °C)
13.68 (up to 300°C) 7.92 (beyond 300°C)	2.1	0.36

The assumed canister is formed by a cylinder 200 cm high and 20 cm in diameter. This cylinder reproduces the same ratio length to width (10/1) considered in USA [2] for future real canisters (H=300 cm, D=30cm). Taking into account the short duration of the experiment, the power of the canister has been assumed constant. If waste aged ten years is considered (Table 1) [3], and the power density is 19 watts/liter, the total power in a canister may be calculated as 1.2 kW. The temperature distribution along the heater axis and perpendicularly to it has been calculated, after 1 year of heating (Figures 1 and 2).

The main results are :

a) the temperature increase 25 m from the heater after 1 year is negligeable ;

b) the thermal gradient is high only in proximity to the heater.

On the basis of these results the decision has been taken to carry out the experiment at a depth of 25 m. Under these conditions the medium may be regarded as infinite. Eight probes for temperature measurements have been placed around the heater at distances ranging from 50 to 200 cm (Figure 3). The clay has been drilled starting from a square cement platform 20 cm thick and with a side of 6 m. Nine boreholes have been drilled with continuous coring. The visual inspection of the cores has indicated a good homogeneity and isotropy of the clay. The boreholes have been cased in the upper three meters to avoid water seepage. The deviations of borehole axis are within tolerance (Figures 4 and 5), showing values varying from 0.25 to 7.00 cm, with a mean value of 2.50 cm. After emplacement of the heater and the probes, the boreholes have been plugged with a mixture of 91 % cement and 9 % bentonite by weight.

A plastic plug, 70 cm thick, has been placed in the borehole 2 meters over the heater, to avoid migration of interstitial water, mobilized by the thermal gradient.

4. INSTRUMENTATION

The heater is shown in Figures 6 and 7. The electrical heater resistance is a pyrotenax cable (Figure 8), consisting of an inner nickel-chrome wire, an insulation coating of magnesium oxide powder and a stainless steel external sheath ; it is fed by alternating current. The heating cable has been helicoidally wound around the inner cylinder of the heater. The heater contains a second identical heating cable for emergency. The inner voids of the heater have been filled with pure quartz sand. The heater is made out of stainless steel (AISI 316). The probes placed in the eight boreholes surrounding the heater are also made out of AISI 316 steel tubing with an external diameter of 5 mm ; this tubing is provided with four openings for the passage of four shielded iron-constantan thermocouples ($\emptyset_{ext}$ = 1.6 mm), as shown in Figure 9. The measurement of the maximum temperature at the heater-clay contact is assured by termocouples placed in three grooves, cut at 120° one from each other in the external wall of the heater. Figure 10 shows the connection scheme of the two heating cables to the electric network. The thermocouples have been connected to an automatic data acquisition system provided with 50 inputs and with a sensitivity of 0.1°C. The output is in the form of tapes printed out at fixed times (Figure 11).

5. RESULTS OF THE EXPERIMENT

The thermocouples have shown pratically constant temperature seven months after the connection of the heater to the network. The measurement interval has been gradually varied from 30 minutes to 6 hours, as the temperature difference between subsequent measurements decreased. The instrumentation has regularly operated during the experiment. The temperatures measured, on the heater surface, have fitted quite well the theoretical values obtained with the Heating 5 code (Figure 12). The thermocouples placed respectively 50 cm over and under the central zone of the heater have recorded temperatures 5 % lower than the central one, in good agreement with calculated values.

On the other hand, the temperatures measured in the clay are anomalous. The temperature increases in the clay recorded in the horizontal plane, including the central point of the heater axis, are reported in Figures 13 to 20. The diagrams indicate many divergences between experimental and theoretical temperatures, with the exception of borehole 9, for which experimental and theoretical temperatures coincide (Figure 20). The thermocouples of boreholes 2, 3, 4 and 5 have recorded temperatures lower than those calculated (Figures 13 to 16), whereas the thermocouples in boreholes 6,7 and 8, the most distant from the heater, have recorded temperatures higher than the theoretical ones. Furthermore these temperatures resulted higher than those measured in boreholes placed near the heater (Figures 17 to 19). The same anomalous pattern was revealed by the thermocouples placed on the other three horizontal planes parallel to the one passing through the center of the heater. Figures 22 and 23 report an example of the evolution with time of the temperatures recorded by the other thermocouples placed in boreholes 7 and 8. Finally the thermocouple embedded in cement, 50 cm above the heater, has recorded a temperature higher than the theoretical one (Figure 21).

In contrast with the theoretical isotropy of the clay formation, heat has shown a preferential diffusion along the direction from the heater to borehole 7. This anomalous temperature distribution in the clay could be explained in one of the following ways :

1. faulty working of the instrumentation ;
2. imperfect parallelism of the boreholes ;
3. anisotropy of the clay surrounding the heater ;
4. seepage of superficial water into the boreholes ;
5. clay perturbation caused by drilling.

Controls have been carried out which exclude malfunction of the instrumentation. The first hypothesis therefore can be excluded.

The verticality of boreholes has been controlled, therefore the second hypothesis can also be excluded.

The examination of the cores from the nine boreholes confirms the substantial homogenity of the clay, which shows no sand layers. This fact invalidates the third hypothesis.

In relation to the fourth hypothesis the intrusion of superficial water cannot be excluded. The seepage could have taken place locally in the boreholes along the contact between the clay and the cement plugs. It could have caused an anomalous heat transmission.

The more probable explanation for the observed anomalies is perhaps provided by the fifth hypothesis. The drilling of boreholes very close together may have caused shear zones and fractures, with consequent migration of the interstitial water along preferential channels.

This latter hypothesis could be confirmed by additional drilling through the experimental zone. Wells should be drilled both in the zone between the heater and the thermocouples and outside the experimental zone. These new boreholes could reveal the existence of induced fractures, water seepage or indicate other possible causes for the observed results of the experiment.

REFERENCES

[1] Heating 5 - An IBM 360 Heat Conduction Program. ORNL/CSD/TM-15. March 1977.

[2] Alternatives for managing wastes from reactors and post-fission operations in the LWR fuel cycle. ERDA 76-43, Volume 2. May 1976.

[3] Siting of fuel reprocessing plants and waste management facilities. ORNL 4451. July 1970.

TABLE I

THERMAL POWER OF HLW.

TIME [a] (days)	POWER [b] OF FISSION PRODUCTS (watts/ton)	POWER [b] OF ACTINIDES (watts/ton)	TOTAL POWER [b] (watts/ton)
90	26200	806	27006
150	19300	646	19946
365	10000	315	10315
3652	1030	71	1101
36525	104	10	114
365250	0,02	2,28	2,30

a) Time after reprocessing.

b) Fission products and actinides present in the wastes generated by the reprocessing of LWR fuel, irradiated to 33000 MWD/TON at 30 MW/TON.

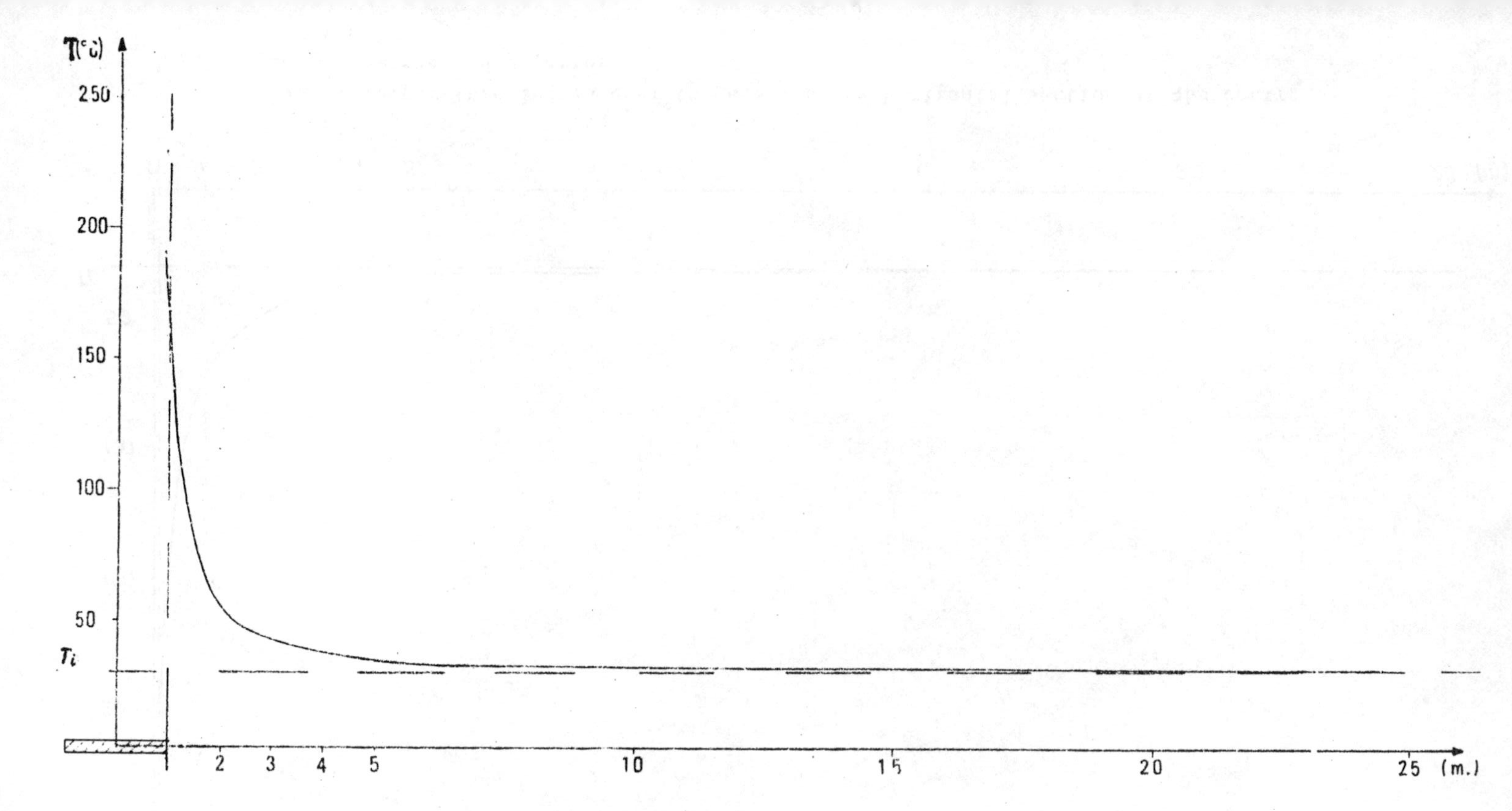

Fig.1. Temperature increase of the clay, along the axis of the heater, after "1" year.

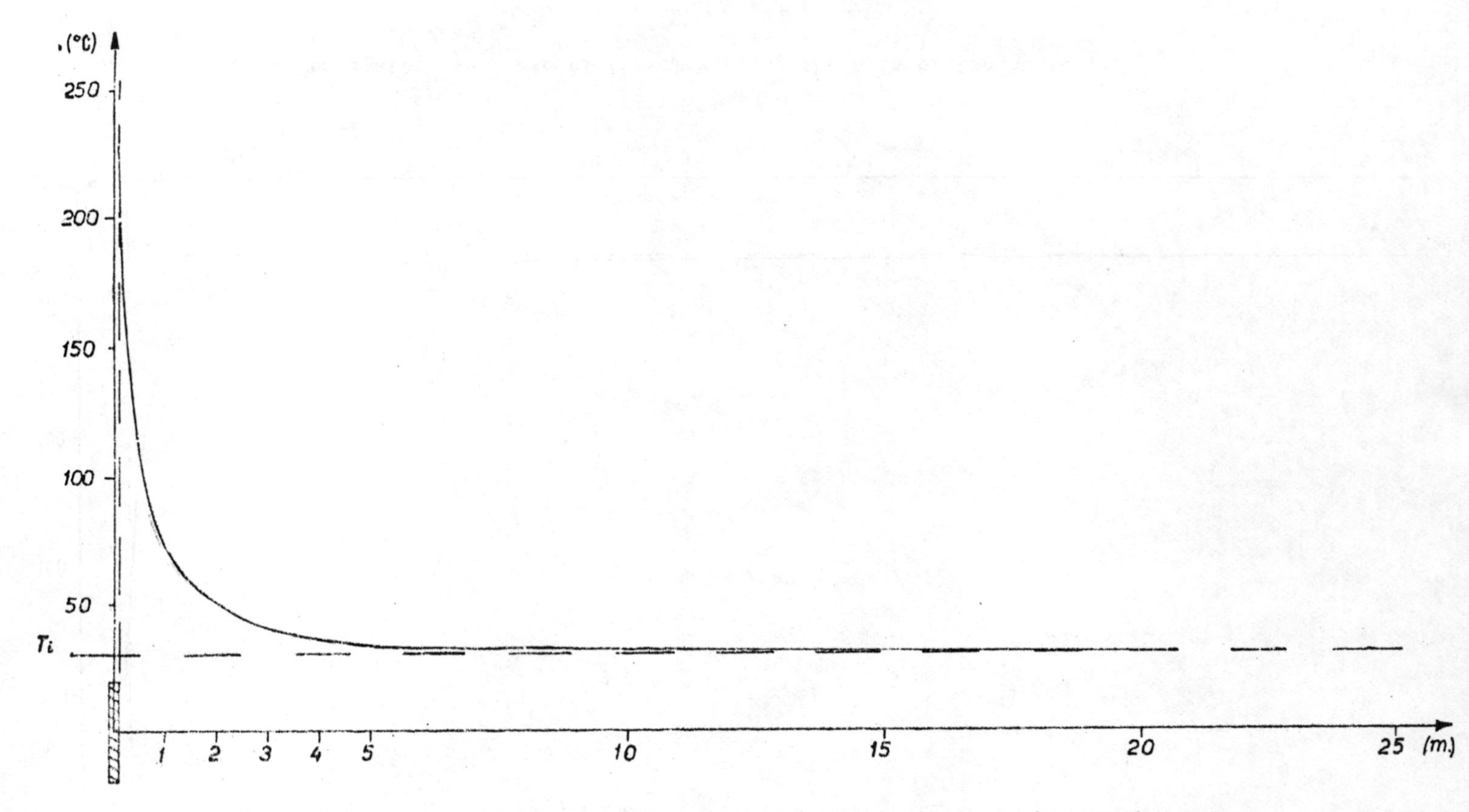

Fig.2. Temperature increase of the clay, on the horizontal section at the centre of the heater, after "1" year.

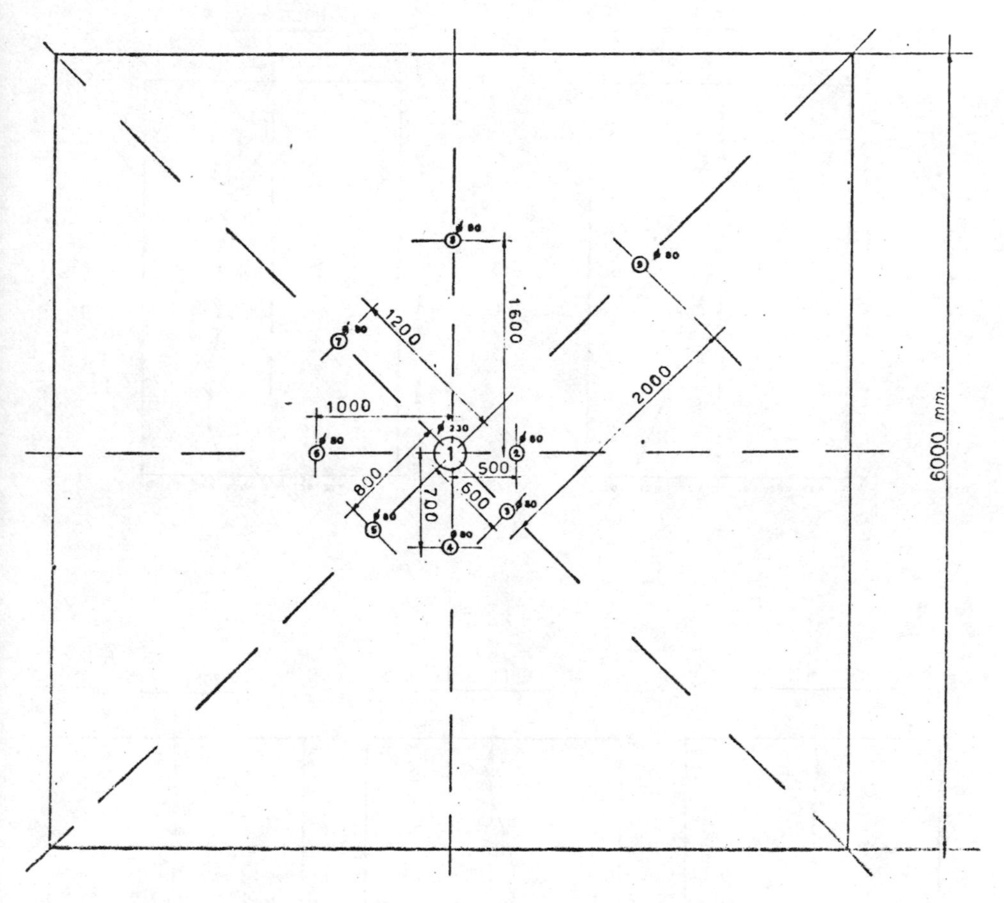

Fig.3. Position of the holes.

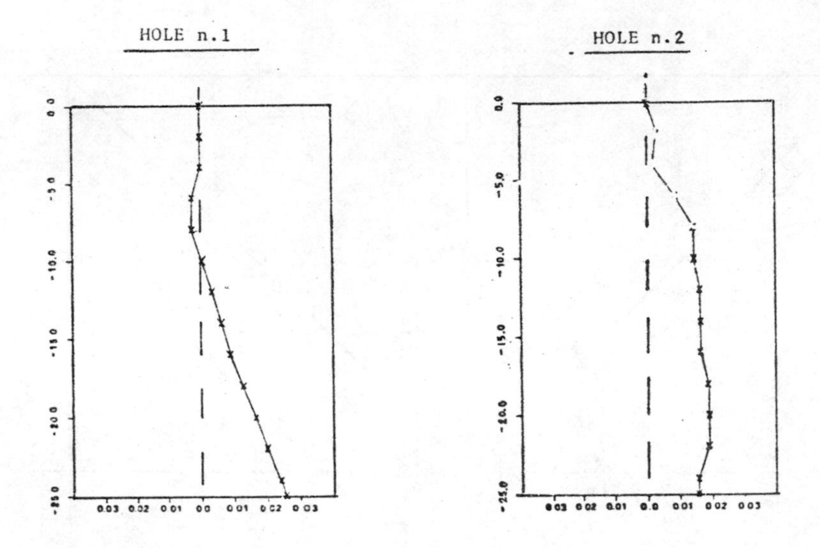

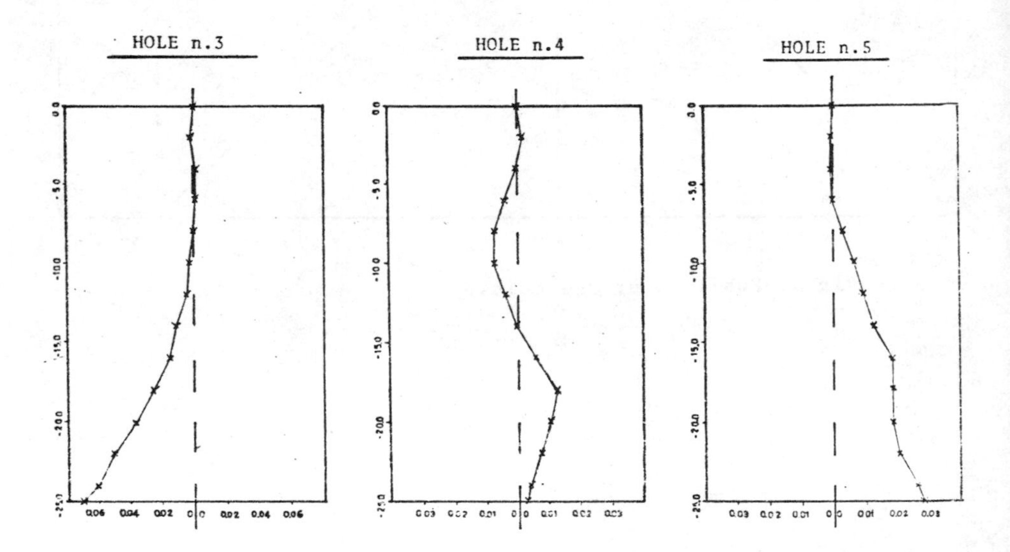

Fig.4. Deviations of the holes n.1,2,3,4 and5.

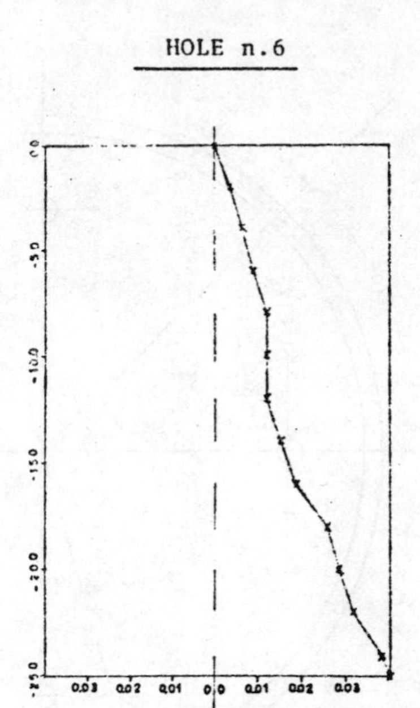

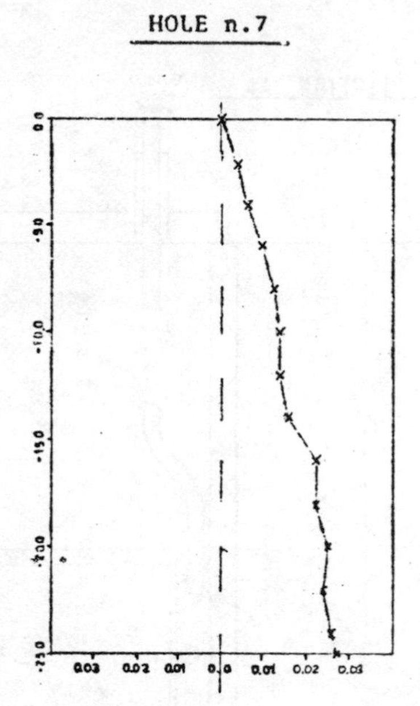

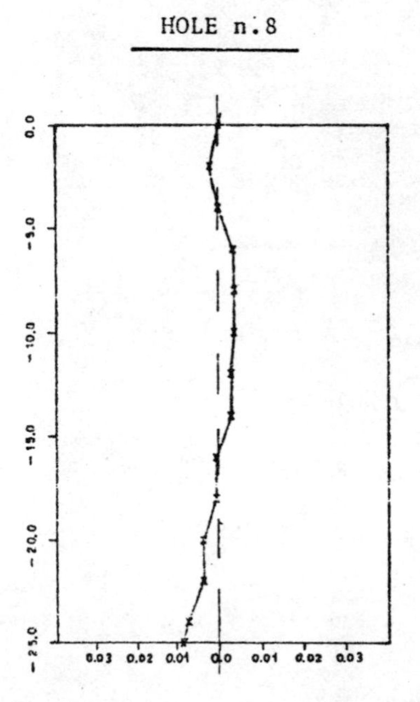

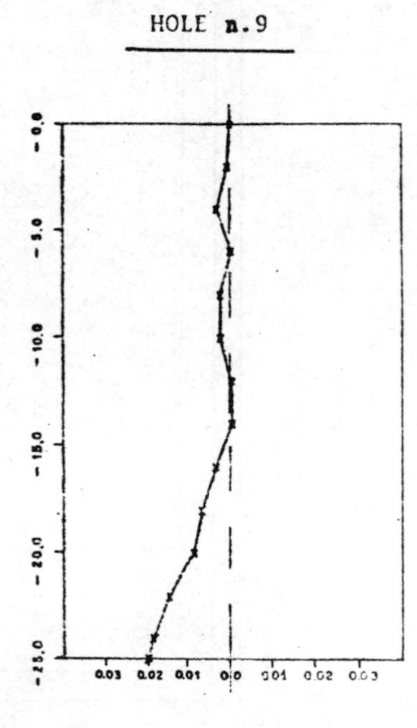

Fig.5. Deviations of the holes n.6,7,8and 9

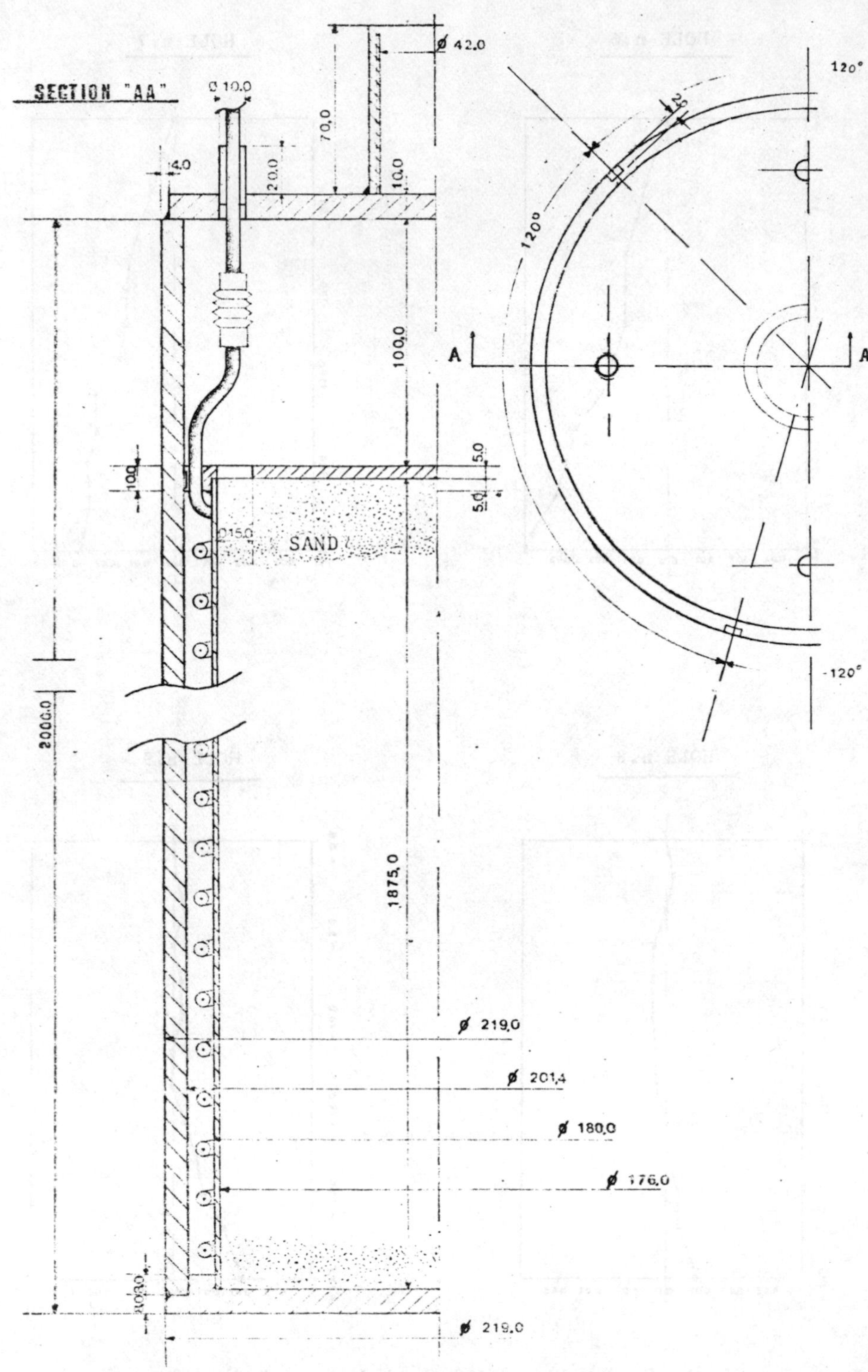

Fig.6. The heater (dimensions in mm.).

Fig.7. Photo of the heater.

Fig.8. Photo of the heating cables.

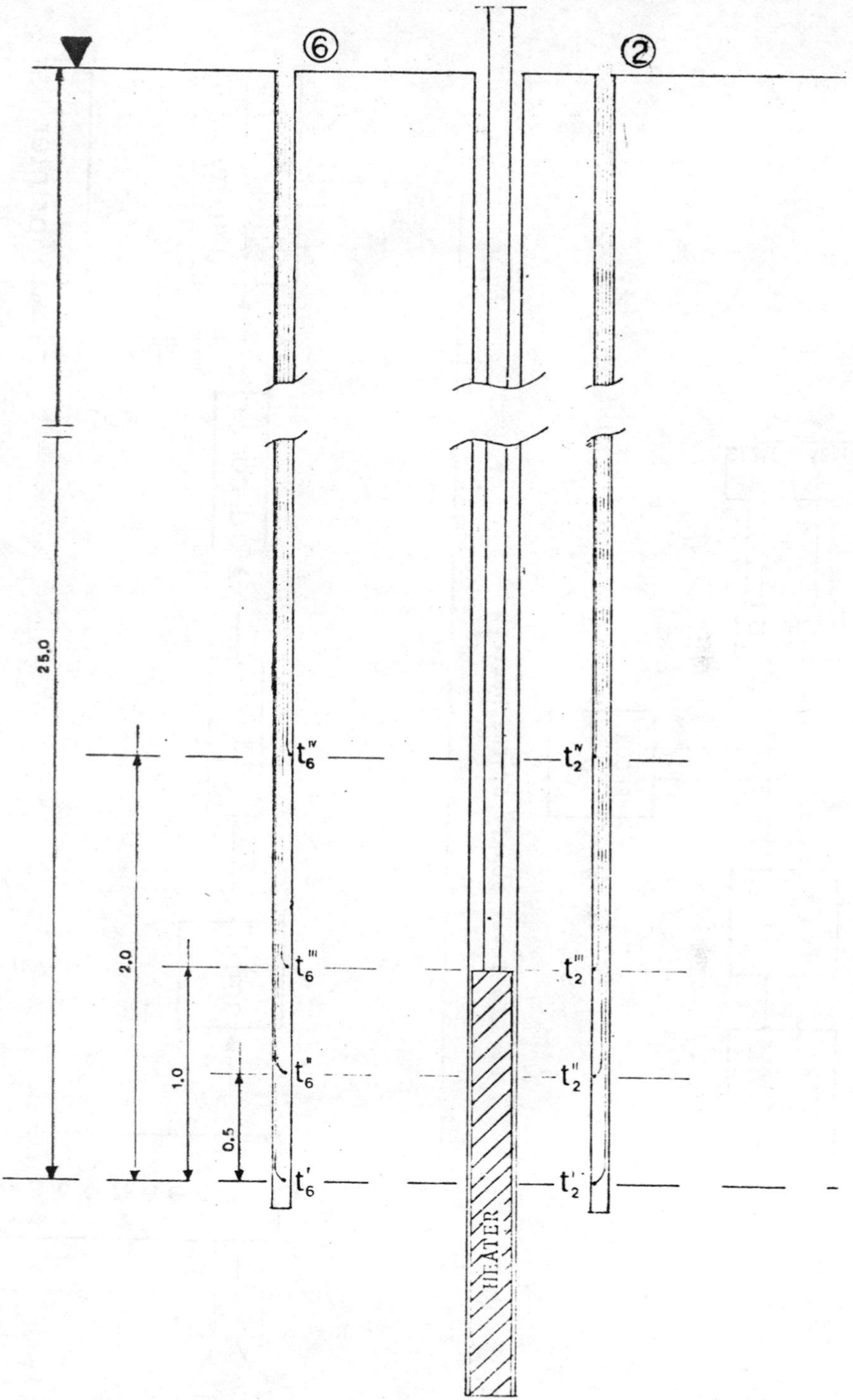

Fig.9. Disposition of the thermocouples in the holes (dimensions in metres).

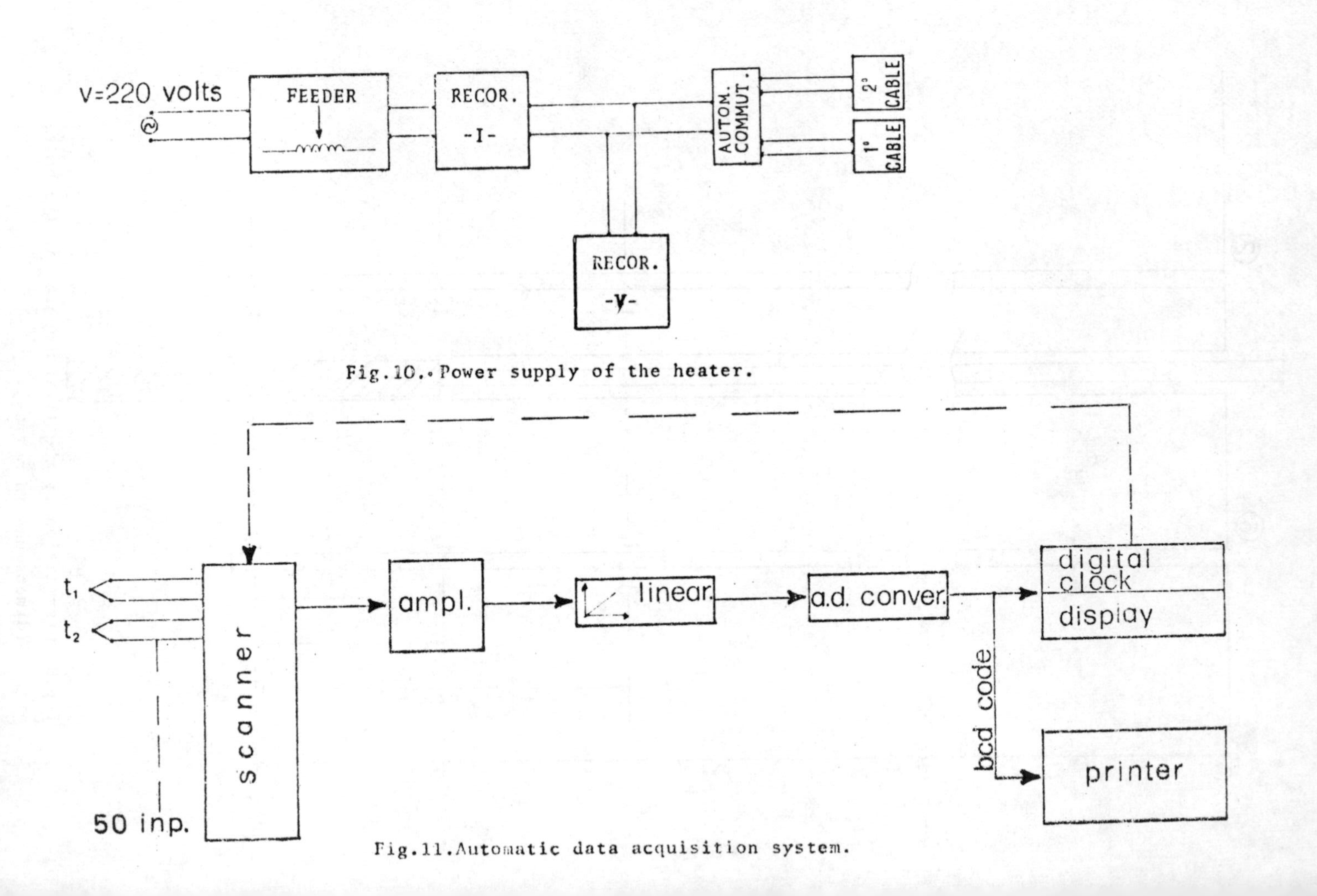

Fig.10.. Power supply of the heater.

Fig.11. Automatic data acquisition system.

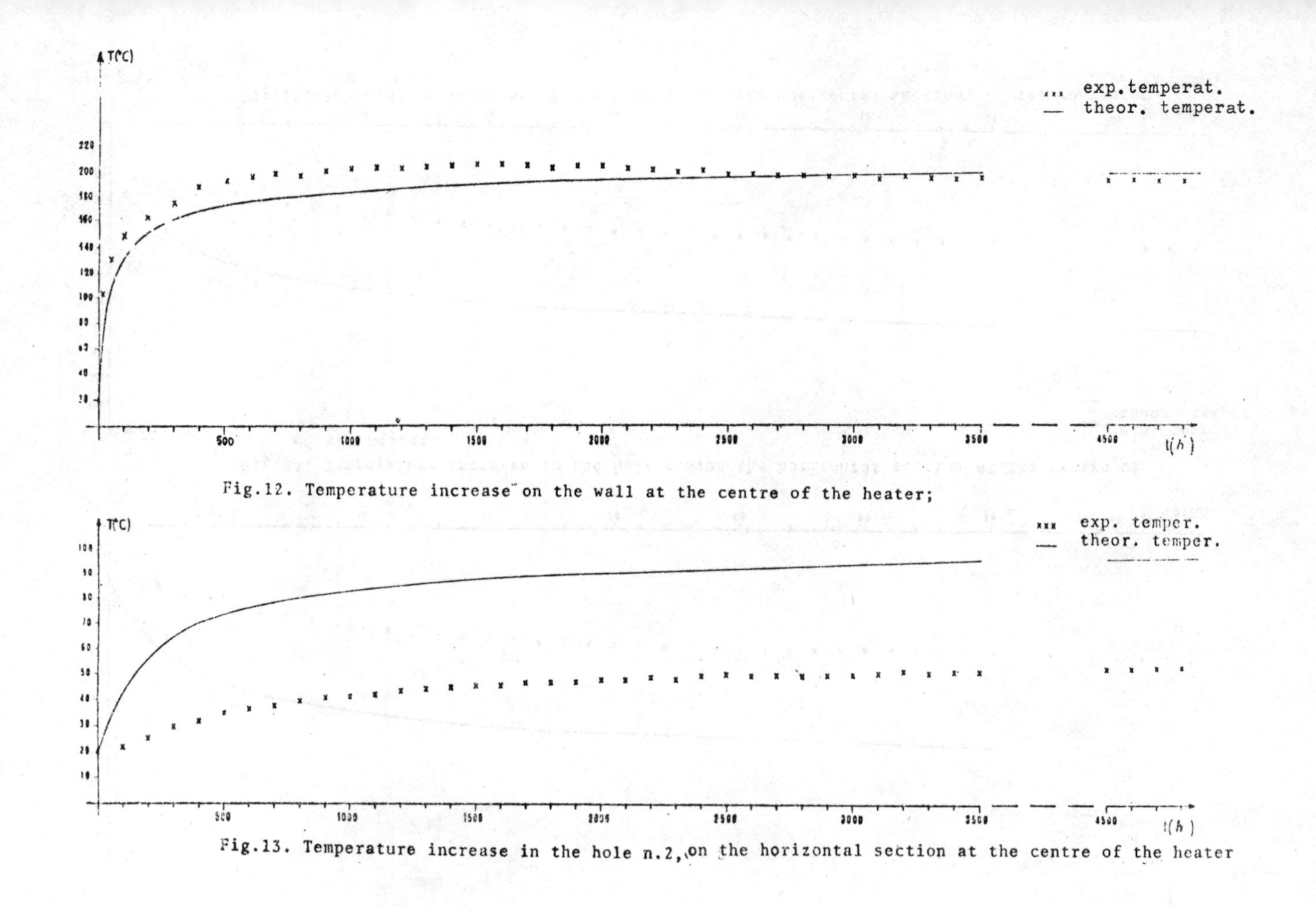

Fig.12. Temperature increase on the wall at the centre of the heater;

Fig.13. Temperature increase in the hole n.2, on the horizontal section at the centre of the heater

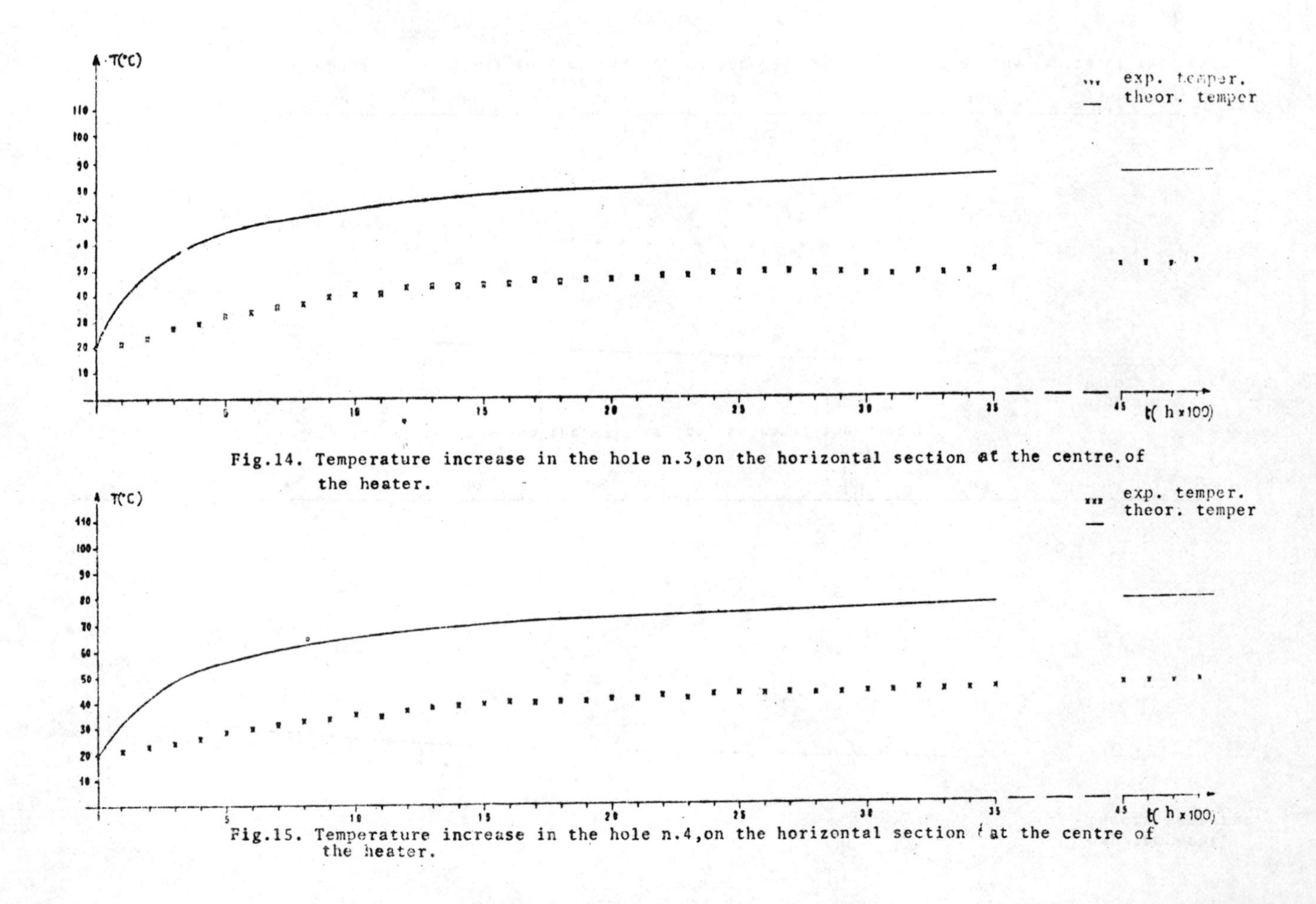

Fig.14. Temperature increase in the hole n.3,on the horizontal section at the centre of the heater.

Fig.15. Temperature increase in the hole n.4,on the horizontal section at the centre of the heater.

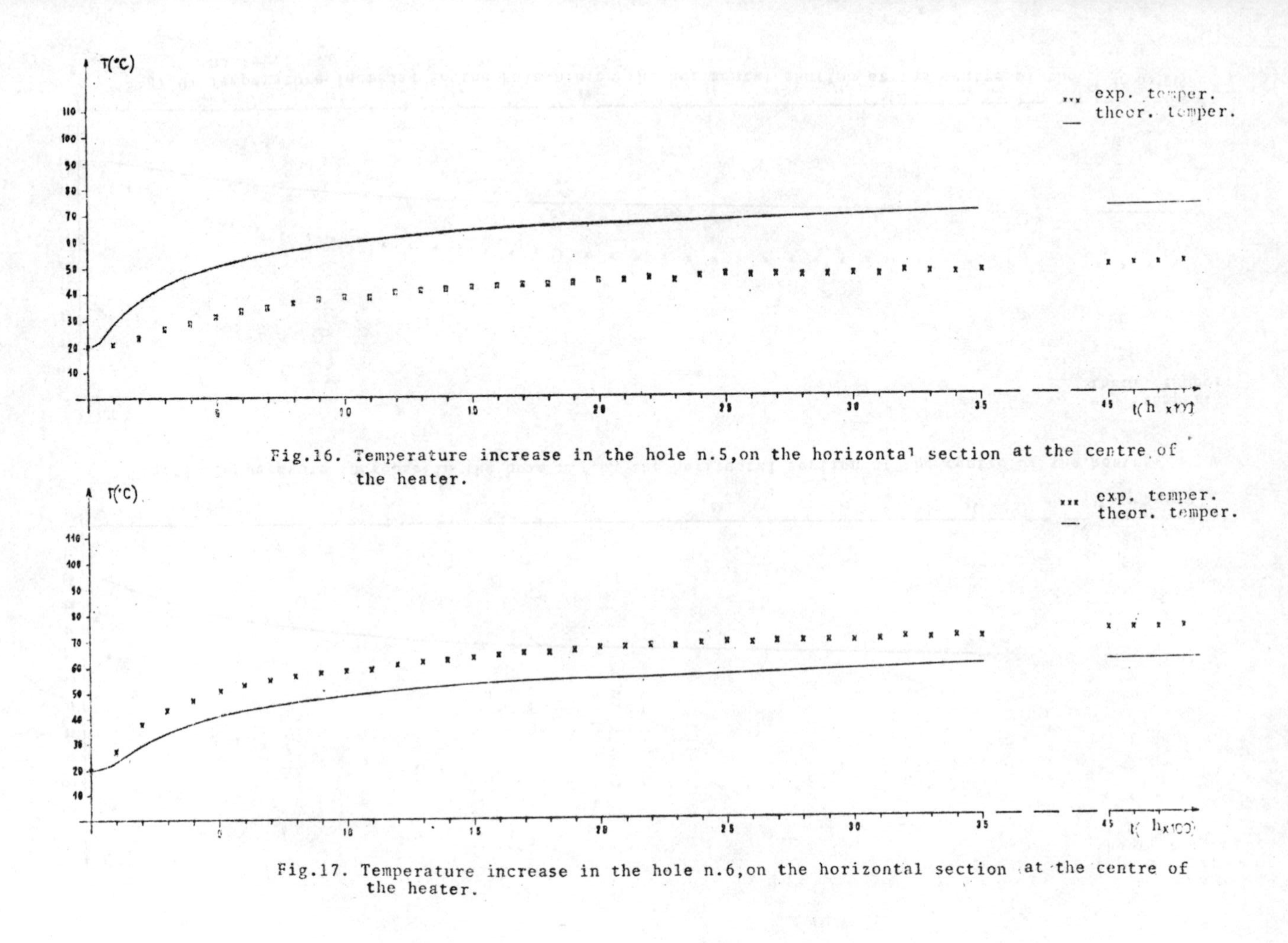

Fig.16. Temperature increase in the hole n.5,on the horizontal section at the centre of the heater.

Fig.17. Temperature increase in the hole n.6,on the horizontal section at the centre of the heater.

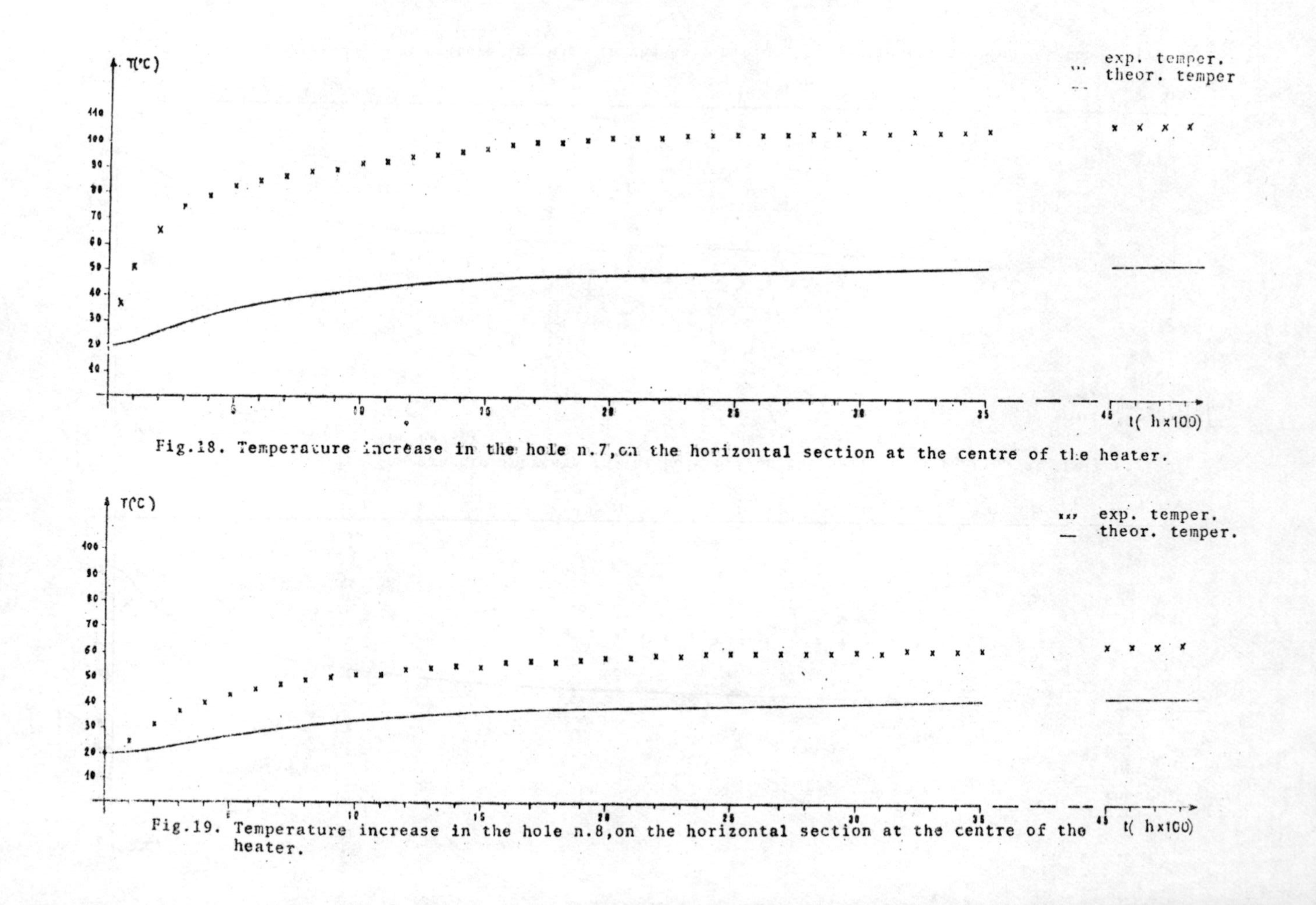

Fig.18. Temperature increase in the hole n.7, on the horizontal section at the centre of the heater.

Fig.19. Temperature increase in the hole n.8, on the horizontal section at the centre of the heater.

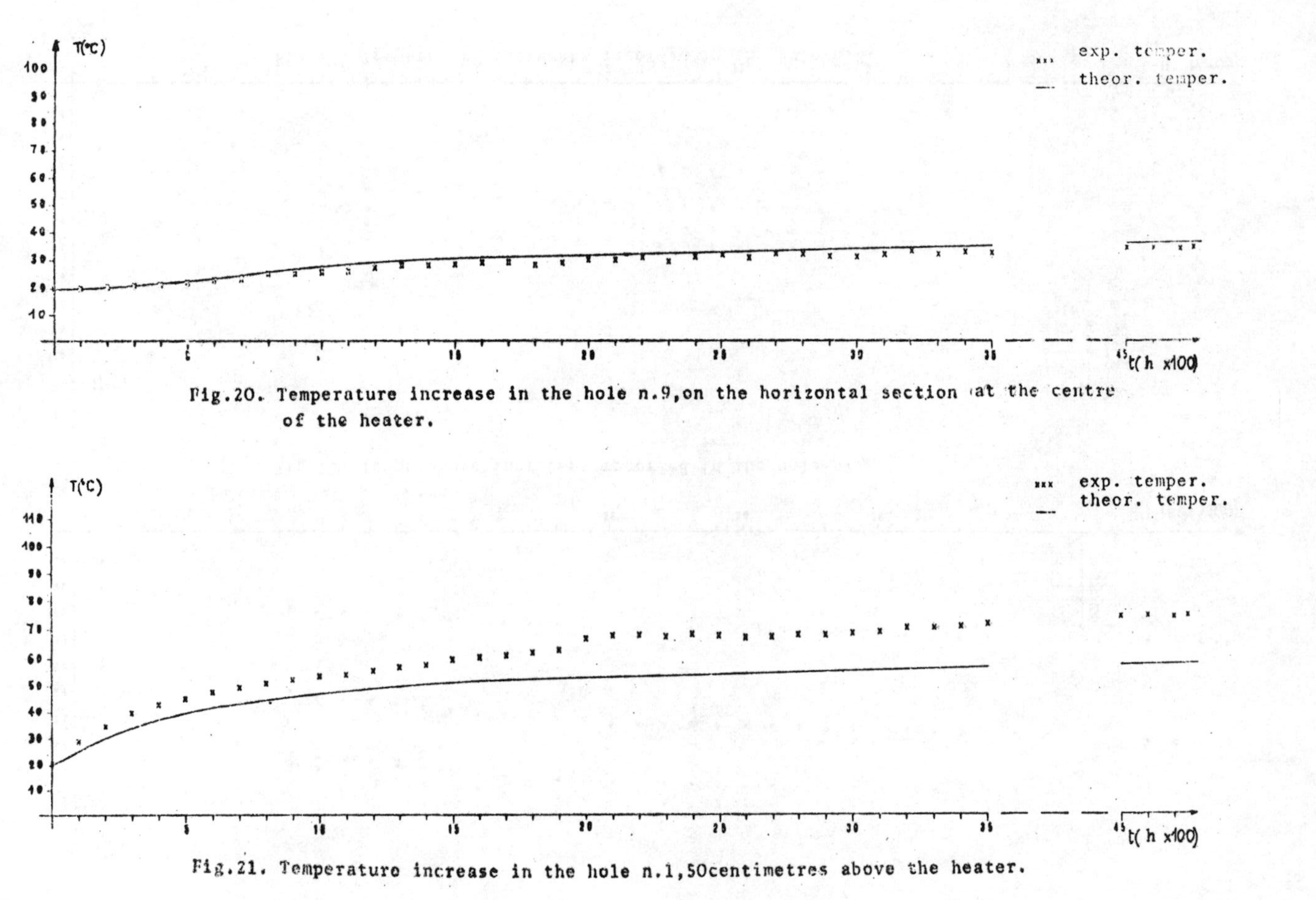

Fig.20. Temperature increase in the hole n.9,on the horizontal section at the centre of the heater.

Fig.21. Temperature increase in the hole n.1,50centimetres above the heater.

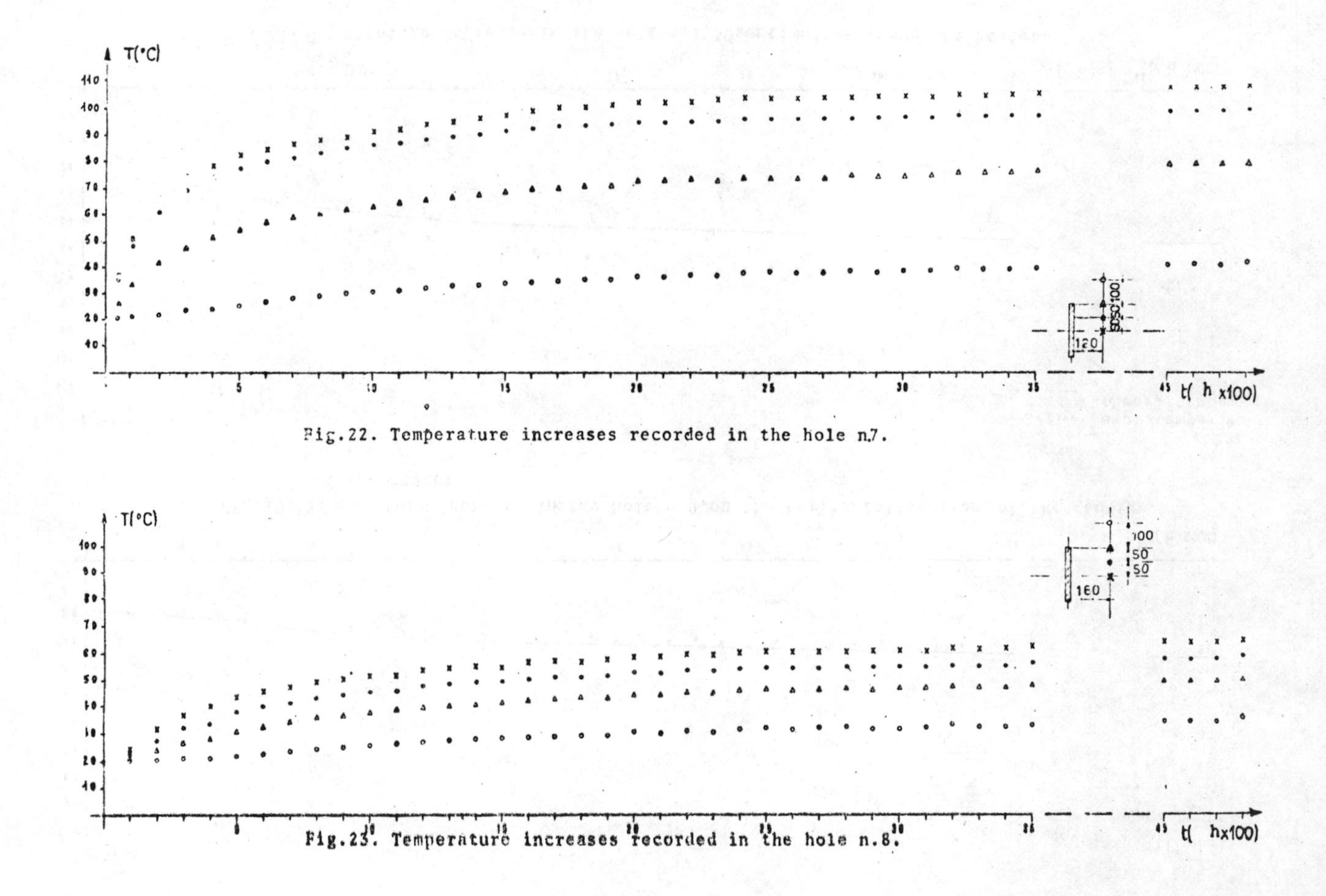

Fig.22. Temperature increases recorded in the hole n.7.

Fig.23. Temperature increases recorded in the hole n.8.

Discussion

J. MARTI, United Kingdom

Besides the visual inspection, have you conducted any tests to verify your assumption of thermal isotropy of the clay ?

E. TASSONI, Italy

No, we have not conducted any additional tests.

J.B. LEWIS, United Kingdom

Are you intending to carry out hydrological studies to examine the possibility of fissuring in the clay ?

E. TASSONI, Italy

No, because we think that fissuring in the clay is entirely due to drilling of the boreholes in the experiment site.

D.F. McVEY, United States

What was the source of the thermal conductivity and specific heat data used in the calculations which were compared with the experiment results ?

E. TASSONI, Italy

The thermal data have been obtained by laboratory measurements carried out at the University of Bari.

D.F. McVEY, United States

Were any attempts made to improve the fit between computations and measurements by changing the thermal properties used in the calculations ? Perhaps conductivity or specific heat were changed by alteration of minerals or dewatering.

E. TASSONI, Italy

We have attempted to improve the fit between experimental and theoretical data by changing the thermal properties of the clay, but we have not obtained a better fit. The changes of thermal properties, during the heating, cannot explain the anomalous experimental data.

L.R. DOLE, United States

Is there a plan to examine the material exposed in situ to heat in this experiment for changes in physico-chemical properties ?

E. TASSONI, Italy

Yes, we think it is necessary to examine the heated material to understand the results of the experiment. Therefore, we have the intention to drill the experimental site and to perform physico-chemical analyses on the core samples.

SOME CONSIDERATIONS ON A BOREHOLE IN A CLAY FORMATION

A. Brondi
Comitato Nazionale per l'Energia Nucleare
Centro di Studi Nucleari della Casaccia
Rome (Italy)

Abstract

A borehole has been drilled in the clay formation underlying the Trisaia Nuclear Research Center (CNEN) in Southern Italy. The local stratigraphic series includes 850 m of marly clay of Pliocene-Calabrian age. The drilling operation has been interrupted at about -400 m, due to the occurrence of methane gas. The presence of gas, the reducing conditions of the clay and the lack of water in fracture zones testify the extremely low permeability of this clay formation. Reducing conditions may prevent the migration of many radionuclides. The occurrence of some sandy levels and lenses is due to the coastal character of the paleosedimentary environment. Previsions on the homogeneity of clay bodies may be indirectly inferred by examination of feeder paleobasins.

Clay is generally considered one of the most suitable geological media for the disposal of nuclear waste. Though other geological formations potentially capable to isolate waste, are distributed in some areas of the Italian territory, clay seems to offer the best solution. Clay is widely spread along the whole Adriatic side of the Italian peninsula and it is present in many more isolated basins included between the Apenninic chain and the western coast. Important clay deposits are also present in Sicily. Italian clay formations are generally distinguished in flyschoid clay of Cenozoic age and Plio-Quaternary clay. Flysch has been often deformed and uplifted by the Tertiary orogeny, the younger clays generally underwent little or no tectonic deformation. This is the main reason why a Plio-Quaternary clay formation has been selected for the initial investigations.

Two complementary lines of research seem to provide the most suitable approach to the selection of appropriate areas and suitable geological situations for waste disposal. They are respectively : wide scale geological surveys and local investigations by means of both field and laboratory work. The equilibrium between these different kinds of research, both aimed at the same goal, represents a significant problem which may vary widely according to the different situations. The necessary studies are intended to provide quantitative information on the attitude and capacity of the geological formations to constitute an effective and lasting barrier to the migration of radionuclides. The quality and quantity of the related analyses must be more and more accurate as the conditions which make the formation valid as a barrier become critical. As an example this may be the case if the repository interferes with important aquifers or if the dimensions of the host formation are not sufficiently large in comparison to the dimensions of the repository. The ideal geological conditions for a repository should assure long-term isolation of the radionuclides without supplementary artificial barriers. In other words the possibility to rely upon the geological barriers rather than on man-made systems is preferable and must be explored with the greatest care. From these considerations the necessity of thorough investigations on the best possibilities offered by the geology in each country appears evident and of primary importance. The aims of the two complementary lines of research above mentioned are universally known and constitute the subject of intense activity in numerous countries. This report is therefore limited to the presentation of what has been done in Italy to investigate the geological aspects of waste disposal.

A general overview of the whole Italian territory has been performed in the frame of a program of the Commission of the European Communities aimed at the compilation of a catalogue of geological formations potentially suitable for the disposal of long lived radioactive wastes. The first stage of this action has been completed and the results will be published in the near future. For the specific investigations a site has been choosen taking into account both the local occurrence of an important clay deposit and the convenience of access. This site, located in Southern Italy, is in fact situated within the Trisaia Nuclear Research Center of CNEN. Therefore, the problems concerned with public acceptance have been minimized. The upper stratigraphic column is formed by about 850 m of marly clay with interbedded micaceous sands in the basal and in the median section. The age of this series is late Pliocene-Calabrian. This argillaceous clay sequence is cut by an erosional surface at the top and it is overlain by sediments of middle and late Pleistocene. These latter sediments are formed by intercalating brown continental sands and yellow-gray marine pelites and gravels. Direct stratigraphic data were available by means of a borehole situated one kilometer from the Trisaia Center and drilled for gas. The local sequence resulted to be formed by clay and sandy clay to a depth of 350 meters, by intercalating sand and clay levels between 350 and 480 meters, whereas the only lithological information provided below this depth was the occurrence of clay down to 850 meters. Saline water was encountered in the sandy levels. Occurrence of methane was indicated in the sandy levels at the depth of 450 meters.

Starting from these preceeding informations a borehole was initially planned to a depth of 760 meters with continuous coring. However the drilling has been interrupted because of an out-coming of gas under high pressure at the depth of 380 meters. In spite of this interruption the core of the first 380 meters will enable to carry out all the necessary analyses and experiments for a satisfactory understanding of the behavior and properties of this clay formation. This also constitutes the basis for possible extrapolations and comparisons with the studies going on in other countries on the same subject. The borehole has been drilled with circulation of bentonitic mud to control the closure of the well walls. Geophysical logs have been performed before casing. Analyses and experiments on the core samples and formation water are now in progress. The lithological data herewith presented have been obtained from a classification carried out in the field during drilling. The general recognized stratigraphic column is here described. Pleistocenic sand and gravel constitute the upper part of the series. A stratigraphic hyatus separates this upper level from the underlying Calabrian series. This results to be generally formed by clay with irregularly distributed thin fine sand lenses, varying between 0.1 and 2 mm in thickness. A true sand level is present from 103 to 118 meters. The clay is rather homogeneous and shows a detritic silty fraction. It is mostly of illitic nature. In the middle and upper part of the sequence the clay presents abundant carbonaceous matter and pyritized fossils. From 190 to 225 meters the clay is affected by thin and regular fractures with an inclination of 45°. It is not yet clear if these fractures are due to some tectonic event or to drilling. It may be supposed that underground the fracturation is only latent. The thin sandy levels are horizontal down to 200 m, at greater depth they show a 20° dip. The thick sand layer contains water with abnormally high pressure. The geochemical analyses indicate that this water is isolated from both phreatic and marine water. The Redox potential is negative. In addition to the main gas flow mentioned above other gaseous manifestations have been encountered at the depths of 100 and 220 meters. The clay plasticity has shown a four-fold variation going from the surface to the deep layers.

Some considerations can be drawn from the geological information obtained at the Trisaia site. Gas bearing layers, the particular geochemistry of the deep water, the lack of water in the fractured zone prove the extremely low permeability of this clay formation. As mentioned above the deep water indicates a reducing environment. The more general reducing character of the clay environment is also testified by the frequent occurrence of organic matter and pyritic nodules. It is generally known that reducing conditions represent one of the most important properties of this geological formation, since it may be an effective barrier to the migration of many radionuclides.

The occurrence of sand levels interbedded with the clay and their evident inclination below 200 meters of depth is probably related to a substantial sublithoral character of the depositional environment. The results of the micropaleontological analyses strongly support this statement. The accumulation of large amounts of argillaceous materials in this particular environment, very near to the source area, has been possible only because of the large occurrence of clay rocks in the feeder basins. Otherwise deposition of pure clay sediments would have been possible only in areas corresponding to a pelagic environment, that is far away from the coast. After the present borehole the stratigraphic column is known in much greater detail than it was previously by means of the borehole situated 1 kilometer from the Trisaia Center. Informations regarding occurrence of water and minor gas-bearing levels were lacking. This is clearly due to the different purposes of the two boreholes. It is evident that information obtained from preexisting boreholes are quite useful but they must be taken into account only as very general indications.

The deep borehole at the Trisaia Center has represented a valuable opportunity to obtain first hand data about a complex geological formation potentially able to isolate long-lived radioactive waste. Some of the conclusions that can be drawn from this experience are the following :

1) Geotechnical investigations in clay are generally limited to some tens of meters from the surface. Therefore little is known about the geotechnical characteristics at great depth. The Trisaia cores should give useful indications on this subject. Efforts have been made to collect undisturbed samples. The Trisaia clay appears to have been overconsolidated by overlying ancient deposits now completely eroded. It therefore offers the opportunity of obtaining geotechnical data on samples corresponding to considerable depth.

2) Thick formations suitable for hosting waste repositories are usually found as filling materials of ancient sedimentary basins, where important reservoirs of ground water and sometimes hydrocarbons can also be found. In many of these basins numerous boreholes have been drilled and their stratigraphy is generally known. Nevertheless the borehole nearby the Trisaia Center shows that the stratigraphic detail provided by these boreholes is not generally sufficient for the characterization of disposal formations. It is therefore necessary to rely on the indirect evidence provided by the petrographic and mineralogical study of the feeder area and by the paleogeographical study of the accumulation environment. Therefore one of the most useful and simple tools is the observation of feeder paleobasins and any outcropping clay deposits preceedingly derived from the same basins. It is not probable that only argillaceous sediments are derived from a granitic or arenaceous basin. If the feeder paleobasin is composed by rocks generating both sand and clay, thick argillaceous sediments may have accumulated only far away from the paleocoast. Therefore it would be pointless to investigate zones corresponding to the paleocoastal environment. Paleogeography is undoubtedly an essential science to be employed in the preliminary search of sites destined to waste disposal.

3) The Trisaia clay sequence shows many indications of reducing environmental conditions. The gray color of the clay already points to a reducing environment. The samples of formation water have resulted characterized by negative values of the Redox potential. As it is generally known from many studies this physico-chemical condition represents an efficient barrier to the migration of many radionuclides. The reducing character of the Trisaia clay may be regarded as a confirmation of a common feature of clay formations. Finally it is evident that if the dimensions of the host clay formation are very large in comparison to those of the repository the geochemical barrier can be quite redundant in respect to the isolation requirements of the waste.

Discussion

C.R. WILSON, United States

What is the Italian Authorities' policy regarding the long term need to protect buried nuclear waste from accidental encroachment ? How much time is felt to be required before the waste no longer presents a radioactive hazard, and what plans are being made for long term governmental supervision of the site ?

A. BRONDI, Italy

It is generally assumed that long-lived wastes need to be isolated from the biosphere for a half million years. For a complete answer to your question I must specify what the Italian Authorities are planning to do on the subject of nuclear waste. A storage facility is planned to allow cooling of nuclear waste to a temperature of about 100°. This will be a man made facility to be used for the storage of waste during a few decades. Afterwords the waste will be disposed of in a deep geological formation. This solution will be definitive and the chosen geological formation must assure isolation without human contribution. It is in fact impossible to plan and to assure human controls for thousands, or even hundreds of years. Maximum effort must be applied in order to minimize the chance of accidental encroachments. The necessary conditions are a sufficient depth, and a lack of interest for this formation as a possible container of primary materials such as water and hydrocarbons. Water is not so interesting if situated at excessive depth. Deep clays cannot be considered as an important resource since great amounts of very easily available clay occurs at the surface. The ratio between the volume of the repository and that of the host formation is very low. This reduces the probability of drilling through the waste if future drilling in the area were to take place.

F. GERA, NEA

The clay formation underlying the Trisaia site contains sand lenses that hold saline water and/or gas methane, however the dimensions of the lenses are too limited to contain commercially interesting quantities of gas. Therefore the presence of gas does not seem sufficient to cause future drilling for hydrocarbons recovery.

I would like to ask if the high pressures observed in the lenses have been used to estimate the large-scale permeability of the clay formation ?

A. BRONDI, Italy

No, because we have been forced to plug the borehole before we could effect measurements and analyses.

T.F. LOMENICK, United States

This question concerns the permeability or impermeability of the clays.

Your studies give indirect evidence of impermeability within the clays. Have you conducted or do you plan to conduct hydraulic tests within the borehole ?

A. BRONDI, Italy

No, because the occurrence of gas under very high pressure has forced us to plug the borehole. We have only carried out an outflow test in a sand level encountered at about hundred meters of depth. The resulting data show obviously a high permeability of this sand level. Geochemical data indicate the isolation of this water body from other hydrological systems such as phreatic or marine ones. This corresponds to an equilibrium situation which could have been perturbed by the drilling. The interruption of the operations has prevented us from verifying if this isolation has been maintained.

B. FEUGA, France

Vous avez mentionné des analyses de l'eau du sondage. Par quel moyen avez-vous prélevé des échantillons d'eau non contaminées par le processus de forage et représentatif d'un niveau donné dans celui-ci ?

A. BRONDI, Italy

Nous avons revêtu les parois du puits et nous avons procédé au lavage pour éloigner la boue bentonitique. La pression de l'eau était tellement élevée, qu'elle a donné une colonne d'eau de six mètres de hauteur au-dessus du sol. Elle a permis la possibilité d'un autolavage en profondeur, ce qui a pour résultat que l'eau recueillie était représentative.

Session 2

Chairman-Président
M. D. RANCON
(France)

Séance 2

PARAMETERS AFFECTING RADIONUCLIDE MIGRATION IN ARGILLACEOUS MEDIA*

B. R. Erdal, B. P. Bayhurst, W. R. Daniels, S. J. DeVilliers,
F. O. Lawrence, J. L. Thompson, and K. Wolfsberg
Los Alamos Scientific Laboratory
Los Alamos, NM 87545, USA

ABSTRACT

A systematic study of some of the parameters that may affect sorption on the Eleana argillite from the Nevada Test Site (NTS) is reported. The nuclides studied were ^{85}Sr, ^{95m}Tc, ^{137}Cs, ^{141}Ce, ^{152}Eu, ^{237}U, ^{237,239}Pu, and ^{241}Am. The parameters studied were time, temperature, cation exchange capacity, available surface area, particle size, radionuclide concentration, groundwater composition, and sampling location within a given structural block. Sorption tends to increase somewhat with time, and desorption is frequently more difficult than sorption. There is no strong correlation between the sorption capability and the sampling location. Particle size and available surface area are not too important. The dependence on temperature depends on the element. Sorption of technetium(VII) and uranium(VI) is low except perhaps when fine sieve fractions are used. The composition of the water seems to be a major factor governing sorption. A proper method for performing static sorption measurements was developed.

*Work performed under the auspices of the U. S. Department of Energy.

1. INTRODUCTION

A major requirement for the evaluation of the long-term safety of nuclear waste disposal in a geologic environment is a thorough understanding of the mechanisms and phenomenology of the sorption-desorption behavior of the various radionuclides that are biologically hazardous. This knowledge will help in the prediction of the fate of the radionuclides during the length of time required for radioactive decay to reduce the waste to safe levels.

This report presents the results of laboratory studies [1] of the sorption behavior of various radionuclides with the Eleana argillite found at the Nevada Test Site (NTS). The effects of some of the variables that can influence sorption and transport properties have been systematically studied using a newly developed static technique. The variables were the available surface area, cation exchange capacity, temperature, time, groundwater composition, radionuclide concentration, and sampling depth within a given structural block.

2. GEOLOGIC MATERIAL PROPERTIES

2.1 Samples and mineralogy

The argillite was obtained from drill hole UE17e in the upper part of Unit J of the Eleana formation at the NTS by A. R. Lappin, Sandia Laboratories, Albuquerque (SLA). The hole is within the Syncline Ridge structural block. The samples studied were from depths of 120 m, 365 m, 548 m, and 717 m. Portions of these samples were pulverized and graded with sieves, and the 106-150 μm and 355-500 μm fractions were selected for study. The fractions were washed briefly with deionized water to remove dust, and they were then dried. Petrographic analyses using thin-sections of these fractions indicated [1] that they contain 25-35% detrital quartz with minor amounts (<4%) of other detrital phases (mostly feldspars) in a ground-mass of hematite (5-9%) and clay minerals. X-ray analyses indicated that the principal clay is montmorillonite with minor amounts of kaolinite. Several other minerals have also been identified [2,3]. The size fractions appeared to have a bimodal distribution of quartz grain sizes in the various fragments. Modal analyses did not indicate any significant mineralogical fractionation with size-sorting. This is consistent with the grain size of individual minerals being much smaller than the smallest fragment. Other portions of these samples were pulverized until the entire quantity passed through a 75-μm sieve. A fourth set of samples was prepared by simply taking a few 2-5 mm diameter pieces of the original sample. The <75-μm and "chunk" samples were also washed and dried.

2.2 Size distribution and surface area measurements

The size distributions of the particles in the 548-m fractions were measured by screening techniques. The median values in Table I were calculated by linear interpolation between screen data that most closely bracketed the median mass. Similarly, the upper and lower quartile values were calculated by linear interpolation between the screen data that immediately bracketed the 75% and 25% mass values, respectively. The semi-interquartile range (SQR) [4] is a measure of the dispersion of the particle size distribution.

---

[1] Erdal, B. R., Aguilar, R. D., Bayhurst, B. P., Oliver, P. Q., Wolfsberg, K.: "Sorption-Desorption Studies on Argillite," Los Alamos Scientific Laboratory report LA-7455-MS (1979).

[2] Hodson, J. N., Hoover, D. L.: "Geology and Lithologic Log for Drill Hole UE17a, Nevada Test Site," U. S. Geological Survey report USGS-1543-1 (1978).

[3] Lin, W." "Measuring the Permeability of Eleana Argillite from Area 17, Nevada Test Site, Using the Transient Method," Lawrence Livermore Laboratory report UCRL-52604 (1978).

[4] Cramer, H.: Mathematical Methods of Statistics (Princeton University Press, 1946), p. 181.

Table I. Particle Size and Surface Area Meaurements

Depth (m)	Mesh Size (μm)	Particle Size Distribution (μm)			Surface Area (m^2/g)	
		Range	Median	SQR*	BET	Gylcol
120	<75**					99
365	<75**					92
548	106-150	75-177	107	12	5.2	48
	250-355					63
	355-500	300-595	415	39	7.6	47
	106-150**	45-210	122	16	12.5	50
	355-500**	250-595	426	38	9.9	64
717	<75**					75

* Semi-interquartile range.
** These samples were not treated by the washing and drying procedure.

The surface areas of the selected fractions were determined by two different techniques, the BET gas adsorption method [5] and the equilibrium ethylene glycol method [6-8]. Nitrogen was used as the adsorbate for the BET measurements. Little difference due to particle size or sampling depth was observed (Table I). The reason for the higher values obtained by use of the ethylene glycol procedure is that the BET method presumably does not include the internal surface area of the clay groundmass.

As expected, the washing pre-treatment did remove the dust generated in the grinding process. However, the effect on surface areas determined using the ethylene glycol procedure was not thought to be significant.

2.3 Cation exchange capacity

The cation exchange capacities of the sieve fractions from the 548-m and 717-m samples were measured [9] using both cesium and strontium. The values indicate that the 548-m samples have a higher specific cation exchange capacity than those from 717-m. The values determined using strontium tended to be higher than those obtained using cesium. However, due to the uncertainties in the method used [9], the differences are not thought to be significant.

---

[5] Brunauer, S., Emmett, P. H., Teller, R.: "Adsorption of Gases in Multimolecular Layers," J. Am. Chem. Soc. 60, 309 (1938).

[6] Dyal, R. S., Hendricks, S. B.: "Total Surface of Clays in Polar Liquids as a Characteristic Index," Soil Sci. 69, 421 (1950).

[7] Bower, C. A., Goertzen, J. O.: "Surface Area of Soils and Clays by an Equalibrium Ethylene Glycol Method," Soil Sci. 87, 289 (1959).

[8] McNeal, B. L.: "Effect of Exchangeable Cations on Glycol Retention by Clay Minerals," Soil Sci. 97, 96 (1964).

[9] Wolfsberg, K.: "Sorption-Desorption Studies of Nevada Test Site Alluvium and Leaching Studies of Nuclear Test Debris," Los Alamos Scientific Laboratory report LA-7216-MS (1978).

Table II. Cation Exchange Capacity Measurements

Sample	(μm)	Cation Exchange Capacity (meq/100g)	
		Cs	Sr
548 m	106-150	14	17
	355-500	14	16
717 m	106-150	8	10
	355-500	8	10

2.4 Groundwater

The water used for the Eleana argillite studies was made up in the laboratory to simulate the composition of a natural groundwater from hole UE16d at the NTS [1,10]. This water is not strictly an Eleana water since virtually all of the production from hole UE16d was from the Tippipah limestone formation overlying the Eleana. Only a small amount of production was from the uppermost quartzites within the Eleana argillite in this hole. This water is therefore representative of waters that would enter the Eleana from above.

Rock-pretreated water was used in all the sorption measurements. This was prepared by contacting batches of the "synthetic" water with pulverized material that had not been sieved. The contact time was at least two weeks with a solution volume to solid ratio of 20 mℓ/g. The phases were separated by centrifugation at 6 000 g followed by filtration through a 0.4-μm Nuclepore filter. This procedure was used for preparation of waters pretreated at ambient temperature (22 ± 2°C) and at elevated temperature (70 ± 1°C). The same rock phase with fresh water was used in all subsequent batches. Detailed chemical analyses of these waters are given in Ref. [1]; typical values (μg/mℓ) are Ba(0.3), Ca(38), Fe(0.02), Li(0.01), Mg(23), K(6.9), Si(16), Na(32), Sr(0.60), HCO_3(190), CO_3(0), Cl(15), F(0.7), and SO_4(70). The pH values were in the range of 8.0-8.6.

3. SORPTION MEASUREMENTS

3.1 Measurement technique

All traced waters used in these studies were prepared using the pretreated waters described previously and carrier-free or high specific activity radionuclides. Several different sets of measurements were made. The isotopes ^{85}Sr, ^{137}Cs, ^{133}Ba, ^{141}Ce, and ^{152}Eu were run as a mixture, as were ^{95m}Tc(VII) and ^{137}Cs. The ^{237}U(VI) was run by itself. The ^{237,239}Pu and ^{241}Am were run separately or as a mixture. Relatively standard radioassay procedures utilizing Ge(Li) and NaI(Tl) detection systems were used [1]. The appropriate volumes of tracers needed for a set of measurements were evaporated to dryness in a washed polyethylene tube overnight on a steam bath. Concentrated hydrochloric acid was added, and the mixture was taken dry again in order to convert the salts to chlorides. Plutonium and americium tracers were dried at room temperature and were not subjected to the hydrochloric acid treatment. The appropriate volume of pretreated groundwater was added, and the mixture was stirred for ~24 h. The mixture was centrifuged for 1 h at 32 000 g, followed by filtration through a 0.4-μm Nuclepore filter. The resulting tracer solution was used for the sorption/measurements within about 0.5 day. The final tracer concentrations were always less than 10^{-6} M.

Batch sorption experiments were performed at ambient temperature and 70°C. One-gram quantities of the crushed rock were shaken with 20 mℓ untraced non-pretreated water for a period of about two weeks. The phases were then separated by centrifuging at 32 000 g for 1 h. The weight of the wash solution remaining

[10] Dosch, R. G., and Lynch, A. W.: "Interaction of Radionuclides with Argillite from the Eleana Formation on the Nevada Test Site," Sandia Laboratories report SAND78-0893 (1979).

with the solid phase was obtained by weighing the tube and solid before and after the pretreatment. A 20-mℓ volume of the tagged pretreated water was then added to the tube, the solid sample was dispersed with vigorous shaking, and the mixture was agitated gently for the selected time. Typically, 1, 2, 4, and 8 week contact times were used. At the end of the shaking period, the aqueous phase was separated from the solids by 4 centrifugings, each in a new polyethylene centrifuge tube, for 1 h at 32 000 g.

The same sorption procedure was also performed using a tube that did not have a solid phase present. This "control" sample was used to indicate if any of the radionuclides were likely to be removed by the container. In all cases, the cesium remained completely in solution. However, this was not the case for most other nuclides studied. It was shown that the amount of sorption on the container varied, depending on whether or not solid material was present, since elements appear to adsorb on any available surface. Therefore, the presence of a solid phase would tend to reduce the fraction of the activity adsorbed on the container. This effect is especially large when crushed rock solid phases are used since they have a surface area appreciably larger than that of the container.

In order to determine the amount of activity remaining with the solid phase, whether due to sorption, precipitation, centrifugation of a colloid with the solid, or by some other mechanism, a fraction (~25%) of the solid was removed for radioactivity assay. The solid phase was well mixed prior to removal of the fraction. The fraction of the solid removed was determined from the activity of ^{137}Cs in the solid aliquot, in the solution, and in the initial solution. This method is reasonable since cesium did not absorb on the container walls. A check was made by weighing the tube before and after removing the sample. In the plutonium and americium studies the entire solid phase was assayed for radioactivity.

Desorption measurements were also made using the radioactively tagged solids from the sorption experiments and fresh rock-equilibrated water. The same experimental method was used.

3.2 Calculations

The equilibrium distribution coefficient, K_d, for the distribution of the radioactivity (activity) between two phases is conventionally defined as:

$$K_d = \frac{\text{activity in solid phase per unit mass of solid}}{\text{activity in solution per unit volume of solution}} .$$

It is not known whether equilibrium is achieved for the types of measurements reported here. However, the distribution of activities between the phases was measured. Therefore, the resulting value is called the sorption ratio, R_d, which is otherwise identical to K_d but does not imply equilibrium.

The following equation was used to calculate the sorption ratios for all cesium, technetium, and uranium analyses, and for most strontium and barium analyses:

$$R_d = \frac{R \cdot A_f - A_t}{A_t} \cdot \frac{V}{W} \qquad (1)$$

where A_f is the radioactivity per mℓ of a given radionuclide in the tagged water (feed) added to the sample, A_t is the radioactivity per mℓ in the supernatant solution after the required contact time, W is the weight (g) of solid material used, V is the total final volume (mℓ) of supernatant solution, and R is the dilution factor which takes the residual solution from the wash into account.

The amount of residual solution left with the solid material was calculated from the weight increase of the sample plus container after the pre-wash, and the measured density of the solutions used. Therefore, $R = V_0/(V_0 + V_r)$ where V_0 is the volume of the tagged solution used.

For those nuclides having a possible problem with sorption on the container (cerium and europium; some strontium and barium), a different calculational method was used. Since a container problem has never been observed for cesium, the sorption ratio for cesium was used as an internal monitor. The activity of the element of interest and of cesium in the solid and liquid samples was measured. The sorption ratio is

$$R_d = \frac{A_s}{A_t} \cdot \frac{1}{W} \tag{2}$$

where A_s is the radioactivity per g of solid. If a ratio of R_d values is calculated using Eq. 2 one has, after rearrangement,

$$R_{dx} = \frac{\left(\frac{A_{sx}}{A_{sm}}\right)}{\left(\frac{A_{tx}}{A_{tm}}\right)} R_{dm} \tag{3}$$

where the x and m refer to the element of interest and cesium, respectively. This equation was used to calculate the sorption ratio for the element of interest since the R_d for cesium was determined in the same experiment and calculated using Eq. 1.

Eq. 2 was used for all sorption and desorption experiments for $^{237,239}Pu$ and ^{241}Am since the activity in the solution and solid was measured directly in these cases.

For the other desorption measurements, the sorption ratios were again calculated assuming that the cesium did not sorb on the container. The radioactivity, A^0_{sm}, of ^{137}Cs on the solid at the beginning of a desorption measurement was calculated using $A^0_{sm} = A^0_m\ (1-f_m)\ (1-f_d)$ where A^0_m is the cesium radioactivity at the beginning of the sorption measurement, f_m is the fraction of the cesium radioactivity remaining in solution after the sorption measurement, and f_d is the fraction of the solid removed from the sample prior to beginning the desorption measurement (obtained from the ^{137}Cs radioactivity on the solid aliquot). The cesium sorption ratio was then calculated by

$$R_{dm} = \frac{A^0_{sm} - A_{tm} \cdot V}{A_{tm} \cdot V} \cdot \frac{V}{(1-f_d)\ W}$$

The sorption ratios for all other species in the desorption measurement were then calculated using the sorption ratio for cesium and Eq. 3.

4. RESULTS AND CONCLUSIONS

Detailed presentation of the results from all of the measurements performed is not within the scope of this report. However, representative values for the sorption ratios are given in Table III. Even though there are some effects of particle size on the sorption ratio, and also an increase in sorption with time, only the average value has been reported. These averages do not include data obtained using the chunk samples.

There are several general observations that can be made concerning the relation of the sorption ratio to the parameters studied. The scatter in the sorption ratios is sometimes larger than the estimated experimental uncertainties, assuming that one could expect a constant or monotonic change with time. This indicates that strictly identical samples or conditions were not always attained.

The sorption ratios tend to increase somewhat with time. This could be due to alteration of the surface mineralogy even at the rather low temperatures involved in these measurements. The alteration phases may have a strong affinity

Table III. Representative Sorption Ratios (mℓ/g)

Element[a]	22 ± 2 °C	70 ± 1 °C
Sr	135 (10)[a]	322 (37)
	126 (14)[b]	268 (36)
Tc(VII)	25 (1)	2.1 (0.1)
	157 (23)	4.8 (1.7)
Cs	1 990 (120)	1 580 (90)
	3 610 (210)	2 680 (140)
Ba	3 920 (710)	13 200 (3 000)
	5 240 (790)	31 300 (6 700)
Ce(III)	41 900 (6 400)	10 800 (2 500)
	86 400 (14 000)	17 400 (3 900)
Eu(III)	36 000 (5 100)	11 400 (3 900)
	89 200 (6 000)	42 700 (7 400)
U(VI)	5 (1)	
	11 (4)	
Pu	4 200 (1 100)	2 100 (220)
	5 700 (1 400)	2 000 (250)
Am	75 000 (15 000)	6 500 (1 900)
	68 000 (8 100)	29 000 (4 300)

[a] Values in parentheses are standard deviations of the means.

[b] The second value listed for each element is that obtained from the desorption measurements.

for cations or they may simply present additional new surfaces for sorption. Alternatively, according to Helgeson [11], the kinetics of the hydrolysis reactions of silicates is controlled by diffusion of the hydrolysis products from the silicate mineral through a surface layer of intermediate reaction products into the bulk solution. The rate of diffusion is then given by the parabolic rate law. The data for argillite are consistent with this diffusion mechanism. The amount of change in the sorption ratio with time is a function of the sorbing element.

The agreement between the late time sorption and desorption R_d values is usually rather good. However, some desorption values are significantly greater (lanthanides and actinides) than those reached by sorption. One can speculate that the observation may be a consequence of the following phenomena. The solution from which sorption takes place may contain different species of the same element (for example, ions of different oxidation states, differently complexed ions, and various degrees of hydration or polymerization). If the exchange between such species is very slow and they exhibit different sorption characteristics, only one species may sorb strongly while the other remains in solution. The nonsorbing species is then absent in the desorption experiment. Diffusion into the solid is also a reasonable explanation for this observation.

There does not seem to be a strong correlation between the sorption capability of the material and the depth at which the geologic samples were collected. There are indications that the 548-m material is a poorer sorber than that from 717 m but the differences are small. The cation exchange capacity of the 548-m material is somewhat higher than that from 717 m (see Table II). This indicates that the Eleana argillite is fairly uniform in its geologic character as a function of depth.

Qualitatively, an increase in the specific surface area is accompanied by an increase in the sorption ratio. However, the changes are not very large since

[11] Helgeson, H. C.: "Kinetics of Mass Transfer Among Silicates and Aqueous Solutions," Geochemica et Cosmochemica Acta 35, 421-69 (1971).

the surface area did not change much with sieve fraction (see Table I). The observed differences may be due entirely to other factors, such as the differences in the mineral composition of the sieve fractions.

Steady state seems to be attained more quickly at the higher temperatures, at least for cesium, strontium, and barium, and essentially the same R_d value is obtained at the two temperatures for each of these three cations. One can expect that alteration or other chemical or geochemical processes would be accelerated at 70°C and that this could lead to increased sorption. On the other hand, the sorption ratios for cerium, europium, plutonium, and americium seem to vary inversely with temperature. Presumably these elements are more soluble at the higher temperature. The technetium(VII) sorption ratios decrease when the temperatures is raised to 70°C.

The technetium(VII) sorption ratios were low (Table III) but definitely greater than zero. Perhaps this is due to the small amount of carbon or organic material in the argillite which can lead to loss of technetium from solution [12]. It is known [13] that reducing conditions can lead to significant loss of technetium from solution due to the formation of technetium(IV). Perhaps the siderite in the argillite [10] or iron introduced during the grinding process can result in reducing conditions in the tightly capped tubes used in this study. The total iron concentration in these samples was about 6% [14].

As expected, uranium(VI) is poorly sorbed on argillite in contact with air at 22°C. This is thought to be due to the rather high carbonate concentration in these waters, which would strongly complex the uranyl ion. For the smaller particle size (<75 μm) there is a decided increase in sorption ratio. The average sorption value obtained using the <75-μm samples was 71.9 ± 7.0 mℓ/g, while that from the sieve fractions was 4.8 ± 0.6 mℓ/g. On desorption the values were 74.8 ± 5.9 mℓ/g and 11.0 ± 4.1 mℓ/g, respectively. This observation is not understood. However, one should note that the <75-μm particle size material is not a sieve fraction. It was prepared by repeated grindings of the material until the entire sample passed through the sieve. The measurements at 70°C were not made.

A factor of 10^7 change in the plutonium concentration (≈1 x 10^{-6} M for ^{239}Pu and ≈2 x 10^{-13} M for ^{237}Pu) made no significant difference in the sorption ratio within the accuracy of the measurements.

It is important to emphasize that the measured sorption ratios for plutonium and americium include effects other than sorption. There may well be differences in the behavior of plutonium or americium even in supposedly identical solutions at pH ≈ 8 to 8.5, e.g., in the degree of polymerization and radiocolloid formation, and hydrolysis resulting in variations in species (including charge) and particle size. Grebenshchikova and Davydov [15] reported that the charge on colloidal Pu(IV) species may be either positive (at low pH values) or negative (at high) and that the isoelectric pH, or point of zero charge, is in the pH region

---

[12] Rai, D., Serne, R. J.: "Solid Phases and Solution Species of Different Elements in Geologic Environments," Battelle Pacific Northwest Laboratories report PNL-2651 (1978).

[13] Bondietti, E. A., Francis, C. W.: "Geologic Migration Potentials of Technetium-99 and Neptunium-237," Science 203, 1337 (1979).

[14] Erdal, B. R. (editor): "Laboratory Studies of Radionuclide Distributions Between Selected Groundwaters and Geologic Media," Los Alamos Scientific Laboratory report LA-7638-PR (1979).

[15] Grebenshchikova, V.I., Davydov, Yu. P: "State of Pu(IV) in the Region of pH = 1.0-12.0 at a Plutonium Concentration of 2·10^{-5} M," Radiokhimiya 7, 191 (1965).

8.0 to 8.5. Polzer and Miner [16] presented a plot of effective charge (due to hydrolysis) of the americium species vs. pH for a 0.1 M $LiClO_4$ solution. Between pH 8.0 and 8.5 the average effective positive charge per atom of americium varied from ≈1.3 to essentially zero. Therefore, large variations in the behavior of both plutonium and americium could be expected in this pH range.

Considerable effort has been made to learn more about the erratic behavior of plutonium and americium in solutions at pH ≈ 8. Details of these studies have been discussed elsewhere [14]. The method of handling the final samples, e.g., centrifuging or filtering, certainly influences the calculated R_d values, probably due to particle size effects. Species effects should not be as dependent on our method of handling but are probably truly variable. Table IV gives representative R_d values showing the effect of filtering the final solutions through 0.4-µm polycarbonate filter membranes. Centrifuging the final solutions would appear to establish a lower limit to the sorption ratio since crushed rock particles and particulates remaining in solutions would tend to lower the calculated sorption ratio. Filtering the solutions would appear to provide a more accurate sorption ratio by removing rock particles and at least defining the particle size for a "solution." The 0.4-µm pore size may be a little large for particles which would be considered transportable in groundwater. However, for large numbers of samples there is a practical limit to how fine a filter can be used. A useful definition of a "solution" may be one with no particles larger than 0.05 µm. The possibility of sorption on the filter membrane must also be considered.

Since the permeability of the argillite is low, it is not surprising that the R_d values for the chunk samples (Table V) are somewhat lower than those obtained from the ground samples (Table II). However, they are not vastly different. The exception to this is the technetium value at 22°C which is appreciably larger. Perhaps the reduced amount of physical alteration of the argillite has resulted in minimal loss of organic constituents or iron(II).

Table IV. Representative Sorption Ratios

	22 ± 2° C	70 ± 1° C
Pu sorption		
Unfiltered	4 200 (1 100)[a]	2 100 (220)
Filtered	37 000 (12 000)	75 000 (38 000)
Pu desorption		
Unfiltered	5 700 (1 400)	2 000 (250)
Filtered	34 000 (3 700)	97 000 (50 000)
Am sorption		
Unfiltered	75 000 (15 000)	6 500 (1 900)
Filtered	300 000 (42 000)	450 000 (120 000)
Am desorption		
Unfiltered	68 000 (8 100)	29 000 (4 300)
Filtered	480 000 (35 000)	$>10^6$

[a] Values in parentheses are standard deviations of the means.

[16] Polzer, W. L., Miner, F. J., "Plutonium and Americium Behavior in the Soil/Water Environment," Battelle Pacific Northwest Laboratories report BNWL-2117 (1976).

Table V. Sorption Ratios (mℓ/g) for the Chunk Samples

Element	22 ± 2° C	70 ± 1° C
Sr	52 (2)[a]	122 (20)
	47 (3)[b]	95 (8)
Tc(VII)	140 (50)	9.6 (0.3)
	560 (130)	35 (3)
Cs	2 590 (1 520)	1 580 (200)
	5 530 (2 300)	2 550 (500)
Ba	360 (120)	1 020 (560)
	680 (150)	1 530 (1 000)
Ce(III)	>15 000	2 840 (2 410)
		3 380 (810)
Eu(III)	>13 000	3 130 (2 280)
		6 680 (2 360)

[a]Values in parentheses are the standard deviations of the means.

[b]The second value listed for each element is that obtained from the desorption measurements.

Table VI. Effect of Pretreatment on Sorption Ratios (mℓ/g)

Element	Pretreatment	No Pretreatment
Sr	188 (23)	436 (76)
Cs	2 540 (150)	3 430 (740)
Ba	7 350 (950)	15 100 (3 700)
Ce(III)	53 800 (19 500)	142 000 (42 000)
Eu(III)	46 800 (19 300)	118 000 (38 000)

[a]Values in parentheses are standard deviations of the mean.

Table VII. Effect of Water Composition on Sorption Ratio (mℓ/g)

Sample	Element	Water I	Water II
548 m	Sr	230 (6)[a]	92 (2)
	Cs	1 460 (80)	1 010 (50)
	Ba	1 030 (30)	400 (10)
	Ce	1 890 (180)	2 760 (390)
	Eu	2 580 (250)	1 990 (170)
717 m	Sr	700 (150)	170 (4)
	Cs	2 390 (20)	1 630 (60)
	Ba	2 300 (90)	700 (15)
	Ce	5 810 (700)	12 400 (1 610)
	Eu	8 340 (1 080)	10 700 (830)

[a]Values in parentheses are the standard deviations of the means.

In order to check briefly on the effect of the water and solid pretreatment, measurements were made in which neither the rock nor the water was pre-equilibrated. Otherwise, exactly the same procedure as described previously was used. The contact times were 1, 2, and 4 weeks and the <75-μm material prepared from the 548-m sample was used. The measurements were run at ambient temperature. The sorption ratios are given in Table VI. The measured R_d values are somewhat higher than those obtained using the same material with pretreated waters. This may be due to entrapment of the cations during the mass transfer processes that are occurring as the system approaches "equilibrium." The non-pretreated data are in reasonable agreement with those obtained by Dosch and Lynch [10] using water having a similar composition.

Since the composition of the groundwater may have a pronounced effect on the sorption of many radionuclides, several experiments to assess this dependence have been performed. The initial composition of the two waters were (in mg/l) Na(10), K(5), Ca(10), Mg(2), SO_4(5), and Cl(4.5) for the dilute water (called water I) and Na(50), K(5), Ca(50), Mg(20), SO_4(70), and Cl(15) for the concentrated water (called water II). The 250-355μm sieve fractions from the 548-m and 717-m materials were used with the same pretreatment and experimental techniques described earlier. The sorption of strontium, cesium, barium, cerium, and europium at about the same concentrations as described above was studied. Only a 28-d contact time was used.

Assuming that other uncontrolled variables are not important, the composition of the water seems to be a major factor governing sorption (Table VII). As expected the sorption ratios for strontium, cesium, and barium decrease when water II is used. This is undoubtedly due to the increased amount of competing ions. For cerium and europium, the sorption ratios increase with increasing ionic strength. This may be due to the enhanced tendency to form radiocolloids as the sulfate concentration increases. The overall agreement with the earlier measurements (Table I) is somewhat poorer than expected. One should also note that the composition of the more concentrated water is somewhat similar to that used earlier. However, no silicate was present in the latter water before the pretreatment.

Attempts were made to identify the individual mineral components responsible for sorption of uranium(VI) and americium by a microautoradiographic technique [17]. The elements were preferentially sorbed on the clay matrix, with small amounts sorbed onto the detrital quartz and secondary calcite.

ACKNOWLEDGEMENTS

The following Los Alamos Scientific Laboratory personnel are acknowledged for the efforts mentioned: A. E. Norris and R. E. Honnell (particle size analyses), R. D. Aguilar, S. Maestas, and P. Q. Oliver (technical assistance), P. A. Elder and M. E. Lark (sample counting and gamma-spectral analyses), and L. M. Wagoner (typing of drafts and final manuscript).

This work was supported in part by the Waste Isolation Safety Assessment Program which is managed by Battelle Memorial Institute under contract with the Department of Energy (DOE), and by the Nevada Nuclear Waste Storage Investigations project managed by the Nevada Operations Office of the DOE.

---

[17] Thompson, J. L., Wolfsberg, K.: "Applicability of Microautoradiography to Sorption Studies," Los Alamos Scientific Laboratory report LA-7609-MS (1979).

Discussion

D. RANCON, France

Les méthodes et les résultats présentés par M. Erdal concordent avec les expériences sur la rétention qui ont été effectuées en France, sans qu'il y ait eu concertation préalable, on peut donc considérer cela comme une confirmation.

Il faut souligner un cas particulier important concernant la sûreté radiologique : le cas du plutonium. Le plutonium est fortement retenu par les minéraux, mais tous les essais en colonnes ont montré qu'il existait une faible fraction du plutonium entraînée par l'eau sans aucune rétention.

L.R. DOLE, United States

Having by now measured a suite of argillaceous samples, can you correlate the results of Kd measurements with specific mineral phases ? In other words would you expect to be able to predict the performance of a whole rock assembly on the basis of the data on the specific mineral constituents ?

B.R. ERDAL, United States

Qualitatively one can make this correlation, particularly when one also uses the results for granite and tuff. However, the possible correlations have not been quantified.

A.R. LAPPIN, United States

In response to the question about numerical mixing of mineral Kd's to estimate whole rock sorption coefficients I wish to point out that in the case of the Eleana argillite, 100 % of the rock reactivity with technetium was found to be due to activated carbon (not pyrolitic graphite) which is less than 1 % in volume. Such a phase, present in very small amounts, is not usually considered in a mixing calculation. Without having done the laboratory experiment, I am sure we could not have calculated correctly the behavior by using clay-mineral Kd's. The clays appear to be irrelevant in this case.

L.R. DOLE, United States

In your observations of the early breakthrough phenomena, what per cent of the column level comes through in the early fraction ?

B.R. ERDAL, United States

There seems to have been a misunderstanding. We find that the peak of the elution curve for the radionuclides studied comes at a much earlier point than predicted using the sorption ratios obtained in the static tests. We have not studied any of the elements for which other workers have seen the "leakage" phenomena to which you refer.

IN SITU GEOCHEMICAL PROPERTIES OF CLAYS SUBJECT TO THERMAL LOADING

N.A. Chapman
Institute of Geological Sciences,
Harwell Laboratory, Harwell, Oxfordshire, U.K.

ABSTRACT

Compositional variation and geochemical environment in an argillaceous unit are a function of age, depth of burial and mode of origin. This paper considers the variation limits likely to be encountered in potential repository host rocks and examines the significance of factors such as porosity, pore-fluid pressure, total fluid content, and major and accessory mineral component behaviours in controlling the geochemical environment in the neighbourhood of a thermally active waste canister. Particular attention is paid to the use of Eh-pH diagrams in assessing corrosion environments and nuclide speciation. The paper outlines the variables which must be considered when endeavouring to interpret such plots (e.g. temperature, concentration, concurrent reactions and probabilities) and uses the behaviour of various iron minerals found in clay deposits under specific conditions to illustrate the complexities.

The overall thermal stability of various clay and accessory minerals is discussed and extended to attempt to predict behaviour under deep repository conditions, using available data on the diagenetic characteristics of clay-rich sediments. The physical behaviour of fluids in plastic clays is considered and methods evaluated for deriving induced geochemical conditions in a thermally active repository. The latter section is particularly related to canister corrosion studies, in situ experiments, and waste dissolution parameters.

Introduction

Two fundamental questions describe the current state of geochemical knowledge surrounding the thermal behaviour of a clay repository over its active life. The first (what is understood about the physical behaviour of pore and interlayer water on heating, and how does its chemical composition change with depth, temperature etc?) has a parallel in the basic geological research carried out over the years on the diagenetic behaviour of argillaceous sediments, particularly in respect to oil-bearing strata, and can in part be approached from this direction. The second (how does one determine realistic fluid compositions under repository conditions and then relate these to canister/waste corrosion behaviour and subsequent nuclide speciation and migration?) is considerably more complex. It can be approached by using available thermodynamic data and attempting to extrapolate widely available standard state geochemical equilibria (e.g. 25°C 1 bar Eh-pH relations of minerals and aqueous solutions) to cover anticipated conditions of temperature, gas partial pressures and component concentrations, and also by attempting in situ control heater experiments in clays to determine derived atmospheric chemistry. Both techniques have their drawbacks but in combination can be used to arrive at reasonable answers as to long-term geochemical behaviour.

The purpose of this paper is to examine some of the available data and techniques covering these two questions and to discuss the behaviour of common clay mineralogies under repository conditions (including the important non-clay accessory minerals which often control and buffer fluid equilibria in an impure clay and hence determine fluid composition). Simple mineral systems are used as examples of how variation in certain parameters can affect stabilities, redox potentials etc. The behaviour of the matrix clay minerals themselves on increasing pressure and temperature is examined and attention is paid to the role of accessory minerals in controlling pore-fluid chemistry.

From the above it should be clear that much of this study is only applicable to the younger plastic clays and shales rather than thermally more mature slates mudstones and phyllites. This is intentional since it is not possible to cover all the category argillaceous rocks in one paper.

The first step is to examine typical juvenile clay mineralogies and physical properties and to consider how these might vary from site to site depending on age and depth.

General Features of Clays and Included Pore Fluids

Young argillaceous deposits (<30 ma old) display several common features: low degree of compaction, high porosity and pore-water content, low degree of clay particle orientation and a variable content of thermally unstable accessory minerals. All these features are simply a function of the depth of burial of the particular unit, and hence the overburden pressures and geothermal gradient to which it has been subjected since deposition.

Porosity decreases continuously with increasing depth of burial from an initial value of up to 80% to about 30% at 500 m, then less rapidly down to a few percent at depths of 4-5 km. The process is irreversible on subsequent uplift of sedimentary units and it is theoretically possible to estimate the age of a unit from its porosity [1]. Consolidated argillaceous units less than about 60 ma old will have porosities in the 20-50% range [2]; for example the Oligocene Boom clay currently under investigation in Belgium has a porosity of 34-44% [3]. Older units (shales, phyllites through to slates etc.) will not only have lower porosities (often in part due to cementitious pore-fillings of calcite and silica) but will have undergone significant mineralogical change owing to the pressure (P) and temperature (T) gradients to which they have been subjected. The processes of diagenesis leading to these changes will in part be replicated locally in a thermally active repository and since T is of more influence than P in affecting pore fluid behaviour (for example a repository T of 100°C is equivalent to a normal burial depth of 2500 m) and mineralogical readjustments, it is possible to compare repository behaviour to PT diagenetic effects.

One of the principle mineralogical changes is the progressive conversion of the simple montmorillonite structure to a mixed layer montmorillonite-illite structure with consequent release of water and increase in clay mineral grain size and fabric. In natural circumstances this transition begins to occur at about 100°C (nominally at 2.5-3 km depth [4]). With increasing T the interlayering becomes less random and the clay structure stabilises. At 100-150°C there will be roughly equal amounts of illite and montmorillonite with possibly some chlorite [5]. The transition affects the pore-fluid chemistry since potassium is removed from this phase, or from detrital feldspar in the deposit. (The role of accessory minerals in a clay unit will be discussed in more detail later.) It should be noted that the expanded montmorillonite layers do maintain their swelling ability to temperatures of 210-230°C or perhaps even higher [5] which may be significant if they are to be used as sealants.

Geochemical changes apart, the most significant effects of the increases in PT regime are those concerning the hydrogeological properties of the clay. Progressive decrease in porosity combined with the fine particle size of a clay induce a series of effects which lead to complex water flow behaviour which is debatably non-Darcyian. The basic factors involved in this departure from simple porous-medium behaviour are principally pore size differentials, extreme flow tortuosity and electrokinetic interactions between the clay minerals and the water, which can cause high fluid viscosity and some element of reverse flow.

In several studies of the apparent non-Darcyian behaviour of clays Olsen (e.g.[6], [7]) concluded that the departures were due in the main to the highly variable and unequal pore sizes encountered in clays. If there is a sufficiently high hydraulic gradient the flow rate is related to it in a linear fashion, but at low potentials there exists a threshold gradient below which no flow can occur. This is particularly true of compacted montmorillonite (the main component of bentonite) which, because of its exceptionally fine grain size, is the principle clay type which does not always adhere to Darcyian flow behaviour.

Permeability values for clays and shales are quite variable (between 8×10^{-4} to 2×10^{-6} md [4]; and related to pore pressure in that the latter value approaches lithostatic load pressures when permeability is less than 10^{-3} md. At these values pore fluid is expelled very slowly during compaction. Pore pressures cannot approach hydrostatic values (as they might locally in the case of fissure flow in crystalline rocks) unless the permeability is in the order of 1 md, which is more the value which would obtain in sands [8]. Actual pore pressures in young clays and shales is naturally a function of the nature of the sedimentary overburden sequence. At depths greater than about 2 km it is usual for pore pressure to approach lithostatic, but because of the inherent low permeability of some clays abnormally high pore pressures can exist at much shallower depths in both clays and entrapped sandy sequences, although never exceeded the lithostatic load (see for example; [9]). Owing to the crystalline shape of clay minerals some degree of horizontal lamination occurs during compaction of clay units on the sub-microscopic scale. Again, the surface electrostatic charge properties come into play and both factors affect the pore size and spacing.

Returning again to the geochemical properties of compacted clays, it is known that the chemistry of the pore fluids changes with increasing pressure in that the concentration of dissolved species decreases with increasing P. This is true of <u>expelled</u> pore fluids [10] and expelled samples do not necessarily represent the static homogenised equilibrium pore fluid composition. The explanation frequently advanced for this effect is once more related to the electrostatic surface forces of the clays. Base exchange clays (e.g. montmorillonite) tend to concentrate pure water close to individual particles by adsorption. Under compression the electrolyte pore solution can be removed but subsequently anions from the clay minerals are blocked from moving out through the water layer by the negative surface charge of the clay particles. Cl^- ions exhibit this decrease in concentration with pressure and generally speaking the salinity of pore fluids in compacted shales is less than that of the surrounding sedimentary units, or in less compacted clay and shale units. At high degrees of compaction chemical gradients exist in the pore fluids which are stable and not subject to diffusion. The effect of anion exclusion discussed above can lead to removal of cations from fluids passing through a semi-permeable clay (to balance charge) and thus operate as a "filtering" mechanism known as ultrafiltration [11].

It can thus be seen that the composition and behaviour of pore fluids in clays is significantly affected by the ambient pressures and temperatures occurring both during formation of a unit and as a result of the emplacement of thermally active wastes which will perturb the natural PT regime at depth. Unless temperatures are very high in a disposal zone it would seem unlikely that there would be any substantial migration of fluid (or local drying out) once the repository has been backfilled with clay. However, the chemistry of the fluids will be significantly affected, and this will affect not only canister/waste corrosion rates but also the behaviour of leached radionuclides which are transported into the clay body. This review has so far concentrated on the more general properties of clays and is now intended to examine in more detail the geochemical behaviour of the pore-fluids resulting from the thermal anomaly of waste emplacement, in particular the relationship between the stability of both the clay minerals and the accessory mineral components and the fluid phase itself. As was noted in the introduction this is a much more complex issue.

Thermal Stability of Argillaceous Units

Generally speaking the clay minerals encountered in prospective repository host units are thermally stable to temperatures in excess of those likely to be imposed by the waste (say 100-150°C) whilst displaying some degree of enhanced dissolution of silica in the alkaline pore fluid environment. The most significant change likely to occur is that already mentioned; the progressive loss of expandable layers in montmorillonite and replacement by illite layers with consequent release of water. This only occurs in the younger clays which have not already undergone deep burial. The mixed-layer montmorillonite-illites formed are stable at repository depths (or at equivalent induced thermal stresses) not exceeding 1000 m to about 180°C [12]. At depths less than this they will transform to a stable illite + chlorite assemblage as temperatures exceed about 200°C. In the United Kingdom some of the 'argillaceous' units currently under investigation for disposal have already equilibrated to this assemblage along the ambient PT gradient.

The effect of such changes on the local canister environment will depend on the pore-fluid pressure and the exact chemistry of the clay minerals involved; for example it is thought that Ca-montmorillorite is likely to be stable to higher temperatures (150°C) than other fully expandable dioctahedral montmorillonites (such as Na-mont.) and hence the formation of illite might not occur [12]. Similarly if permeability is relatively high such that pore pressure < lithostatic the thermal stability of the clay phases is reduced. If the transition were to occur locally the combined effect of water release and clay volume change would lead to perturbations in both the stress field and pore-fluid chemistry in the immediate neighbourhood of a canister. This behaviour must be correlated with that of other unstable accessory minerals which might occur in a young plastic clay, and which will influence the chemistry of the pore fluids.

The chief minerals of concern are carbonates (of iron, calcium and magnesium), sulphides and oxides (principally of iron) and organic chemicals. In addition minor amounts of sulphates may occur. In clays and shales the stability of carbonates, sulphides and oxides is a function of the Eh and pH of the pore-fluids in equilibrium with them. Alterations in the PT environment caused by the waste canisters will naturally disturb these equilibria such that the Eh-Ph values of the pore-fluids change from those found in situ in the deposit. The volumetric extent of such changes in a repository is unknown at present but is of great importance in modelling leached nuclide behaviour.

Eh-pH Conditions and Their Significance

At present the speciation of radionuclides in ground waters and pore-fluids is modelled using Eh-pH diagrams which are calculated for standard state conditions (25°C, 1 atmosphere) and for given elemental or complex concentrations. Brookins [13,14] has used this approach to interpolate the migration of nuclides in the Oklo natural reactor (in shales) and in doing so has defended the use of standard state data on the basis that small changes in T (in the order of tens of degrees) and P have little affect on the inferred stability fields of the various species. This approach must be treated with caution, particularly when stability fields are compared with assumed limiting Eh-pH boundaries to the environment under study, since most data in the latter area are only applicable to near-surface waters (e.g. [15]). The effects of increasing T on mineral stability fields is in fact small at low temperatures but as T exceeds 25°C by a large factor, as may be the case in a repository, the phase boundaries may shift considerably. Similarly the effect of increasing pressure is only small when the partial pressures of the reacting gases are very low. At higher gas fugacities this is not the case [16]. In addition any particular Eh-pH equilibrium stability diagram is only applicable for given concentrations of species in solution and whilst this is not a great problem when examining nuclide speciation (where the concentration in the leachate groundwater will be vanishingly small) it must be taken into account when modelling stability of argillaceous units, particularly in corrosion studies. Taking any of these factors it is possible to construct three-dimensional Eh-pH-T, Eh-pH-C, etc. diagrams to cover, in part, natural variations. Generally speaking, however, the shape of species stability fields in Eh-pH space remains much the same, only the boundary positions move. This is illustrated in Fig. 1 which shows the standard state stability fields of common iron minerals found in argillaceous units, and the boundary shifts concurrent on changing dissolved iron content. Note, however, that this is for fixed activities of dissolved carbonate and sulphur. Since pore fluids in a plastic clay or shale are likely to be both quite basic (pH8 or greater) and somewhat reducing (Eh < 0) it is likely that siderite, haematite and pyrite will be stable, but it would require only slight PT perturbations to induce breakdown of a given phase such that the activities of gaseous species in pore solution changed progressively. In this particular diagram the change in stability space is best illustrated by the behaviour of pyrite, which under the a_s conditions shown has a relatively small field. However, if a_{CO_3} is reduced to the benefit of a_s the pyrite field expands to occupy most of the reducing area of the diagram [17] and it in fact requires only a very low partial pressure of S_2 to achieve this.

A further feature which must be borne in mind when attempting to apply these diagrams is that they take no account of reaction kinetics, metastability, polymorphism or phase purity and thus the existence of a stability field (for a mineral or even a nuclide species) may not necessarily mean that that phase will be found naturally under given conditions.

All of these considerations have a bearing on both theoretical and experimental studies of canister corrosion behaviour. It is clear that experimental studies which take account only of expelled cold pore-fluid pH and chemistry will only represent one end-member of the corrosion spectrum, and that in-situ values of Eh and pH of warm pore waters might be extremely variable and dependent entirely on the local mineral chemistry. The presence of any quantity of organic material in a deposit will be favoured by reducing conditions and will be involved in buffering Eh accordingly.

Central Geochemical Issues

A number of points can be deduced from the foregoing discussion which will have considerable bearing on the canister geochemical environment and the behaviour of leached radionuclides, when a thermal load is placed on the host clay or shale.

1. There is an inherent instability problem in some montmorillonite dominated young clays which may be encountered if temperatures exceed 100°C and which could lead to fluid release and volumetric changes (evidenced presumably in stress deviations) in the clay around a waste canister.

2. Generally speaking the older argillaceous formations will be thermally more stable in all geochemical respects.

3. The geochemistry of warm pore-fluids in a sealed repository is extremely complex both on a large scale and in individual pore volumes. The geochemical behaviour is governed by four principle factors:

   (a) Eh-pH buffering by component minerals.

   (b) Electrochemical surface forces on individual clay particles leading to ion-blocking and small-scale fluid compositional gradients.

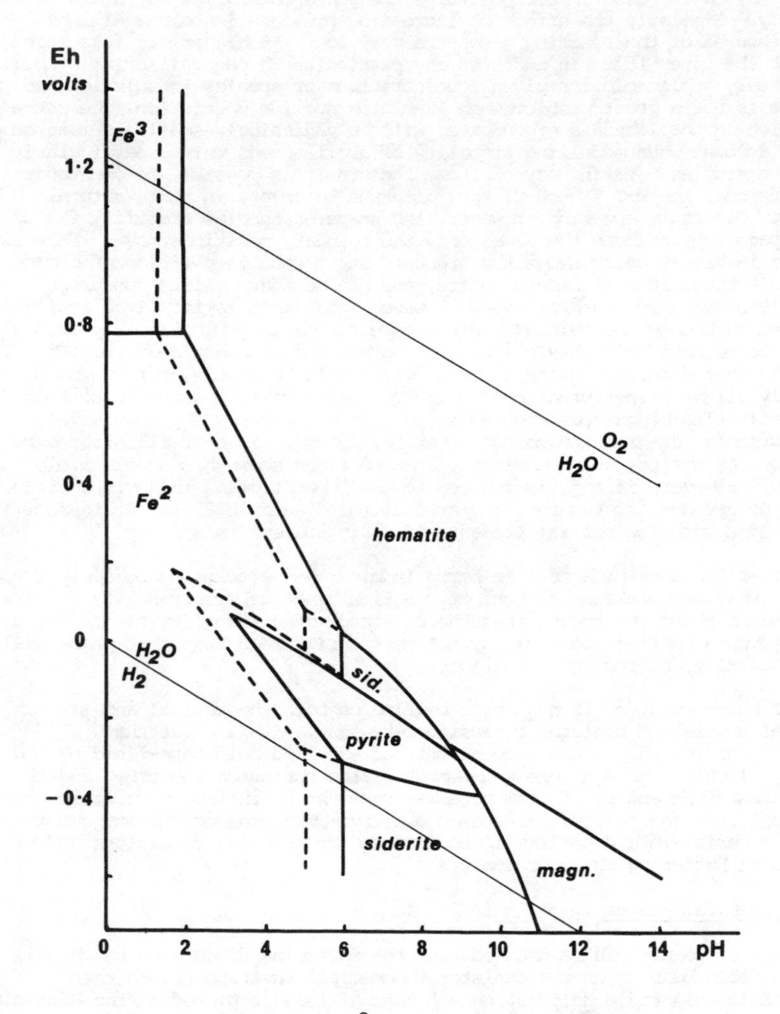

FIG 1. Eh-pH diagram (25°C, 1 atmosphere) showing stability fields of common iron minerals which may occur in argillaceous units. Dissolved component activities: carbonate, 1M; sulphur 10^{-6}M; iron (solid lines) 10^{-6}M; (dashed lines) 10^{-4}M. After Krauskopf (1967).

(c) Ultrafiltration (or membrane filtration) leading to selective retardation of transported species.

(d) Large-scale hydraulic gradient factors affecting speed of movement of pore-fluids and the magnitude of pore pressures.

4. The hydrogeological evidence suggests that local drying out of the clay host is not likely to occur in a sealed repository but may be a problem during the emplacement phase (see for example; [18]).

If one considers these points it is clear that the in situ thermal behaviour must be studied rather than relying on simple laboratory techniques. In the corrosion field this is particularly important and emphasis must be placed on determining the activities of corrosive species in the immediate environment of a hot, emplaced canister. This approach must also be taken when modelling nuclide migration behaviour and it is particularly important to know both the ambient pore-fluid chemistry in undisturbed clay as well as the excursions in composition owing to thermal effects. Considerable caution must be exercised in applying standard state speciation data to the repository itself. In situ migration studies would appear to be the only reliable technique in this context but are fraught with problems owing to the extremely low flow rate in plastic clays.

Clay minerals are exceptionally complex in their behaviour, particularly when in contact with dilute solutions. Of all the rock types currently being studied for waste disposal they are probably the least well understood and the problems associated with their in situ geochemical behaviour the most intractable. However, their intrinsic hydrogeological properties make them amongst the most attractive in terms of integrity and nuclide retardation. Since they are also likely to be used in any proposed hard-rock repository as a sealant and buffer, and since they are the host unit for ocean-bed disposal concepts they must become central to HLW disposal research. In this respect the importance of carrying out in situ geochemical experiments is emphasised.

Acknowledgements

This study was performed by the Institute of Geological Sciences under contract to the United Kingdom Atomic Energy Authority and is published by permission of the UKAEA and the Director, IGS.

References

[1] Muller, G.: "Diagenesis in Argillaceous Sediments", in Larsen, G. and Chilingar, G.V. "Diagenesis in Sediments", Developments in Sedimentology, 8, 127-177. Elsevier, Amsterdam (1967)

[2] Manger, G.E.: "Porosity and Bulk Density of Sedimentary Rocks", U.S. Geol. Survey Bull. 114-E (1963).

[3] Bonne, A.; Heremans, R.; Manfroy, P.; Dejonghe, P.: "Investigations enterprises pour preciser les caracteristiques du site argileux de Mol comme lieu de rejet souterrain pour les dechets radioactifs solidifies", Proceedings of IAEA/OECD Symposium on the Underground Disposal of Radioactive Wastes; IAEA-SM-243 (Helsinki 1979, in press).

[4] Dickey, P.A.: "Migration of Interstitial Water in Sediments and the Concentration of Petroleum and Useful Minerals", Proceedings 24th Int. Geol. Cong., 5, 3-16 (1972).

[5] Weaver, C.E. and Beck, K.C.: "Clay Water Diagenesis During Burial: How Mud Becomes Gneiss", Geol. Soc. Amer. Special Paper 134 (1971).

[6] Olsen, H.W.: "Hydraulic Flow Through Saturated Clays", Proc. 9th Nat. Conf. on Clays and Clay Minerals, Lafayette, 131-161 (1960).

[7] Olsen, H.W.: "Darcy's Law in Saturated Kaolinite", Water Resources Research, 2, 287-295 (1966)

[8] Bredehoeft, J.D. and Hanshaw, B.B.: "On the Maintenance of Anomalous Fluid Pressures: I. Thick Sedimentary Sequences", Geol. Soc. Amer. Bull., 79, 1097-1106 (1968).

[9] Fertl, W.H. (ed): "Abnormal Formation Pressures", Developments in Petroleum Science 2, Elsevier, Amsterdam (1976).

[10] Chilingarian, G.V. and Rieke, H.H.: "Compaction of Argillaceous Sediments", in Developments in Petroleum Science 2, Elsevier, Amsterdam, 49-100 (1976).

[11] Berner, R.A.: "Principles of Chemical Sedimentology", McGraw Hill, New York (1971).

[12] Velde, B.: "Clays and Clay Minerals in Natural and Synthetic Systems", Developments in Sedimentology 21, Elsevier, Amsterdam (1977).

[13] Brookins, D.G.: "Retention of Transuranic and Actinide Elements and Bismuth at the Oklo Natural Reactor, Gabon: Application of Eh-pH Diagrams", Chem. Geol., 23, 309-323 (1978).

[14] Brookins, D.G.: "Eh-pH Diagrams for Elements from Z = 40 to Z = 52: Application to the Oklo Natural Reactor, Gabon", Chem. Geol., 23, 325-342 (1978).

[15] Baas Becking, L.G.M.; Kaplan, I.R.; Moore, D.: "Limits of the Natural Environment in Terms of pH and Oxidation-Reduction Potentials", J. Geol., 68, 243-284 (1960).

[16] Garrels, R.M. and Christ, C.L.: "Solutions, Minerals, and Equilibria", Harper and Row, New York (1965).

[17] Krauskopf, K.B.: "Introduction to Geochemistry", McGraw Hill, New York (1967).

[18] Pusch, R.: "Water Uptake in a Bentonite Buffer Mass. A Model Study", Kärnsbränslesäkerhet Teknisk Rapport, KBS-23, Stockholm (1977).

MATERIAL PROPERTIES OF ELEANA ARGILLITE--EXTRAPOLATION TO OTHER ARGILLACEOUS ROCKS, AND IMPLICATIONS FOR WASTE MANAGEMENT*

A. R. Lappin and W. A. Olsson
Sandia Laboratories**
Albuquerque, New Mexico 87185

ABSTRACT

Results of a near-surface heater test in the Eleana argillite suggest the possibility that the high-temperature (> 100°C) thermomechanical response of argillite to waste emplacement may be dominated by behavior of expandable clays. Enough expandable clay is probably present in most argillaceous rocks to cause a similar response. In situ thermal conductivities may be markedly reduced by even small amounts of clay contraction which results in opening of pre-existing joints. A simple model predicts that such behavior may continue to operate to considerable depths, though several factors affecting determination of this depth remain poorly defined at present.

---

* This work was supported by the U.S. Department of Energy, DOE, under contract DE-AC04-76DP00789.

** A U.S. DOE facility.

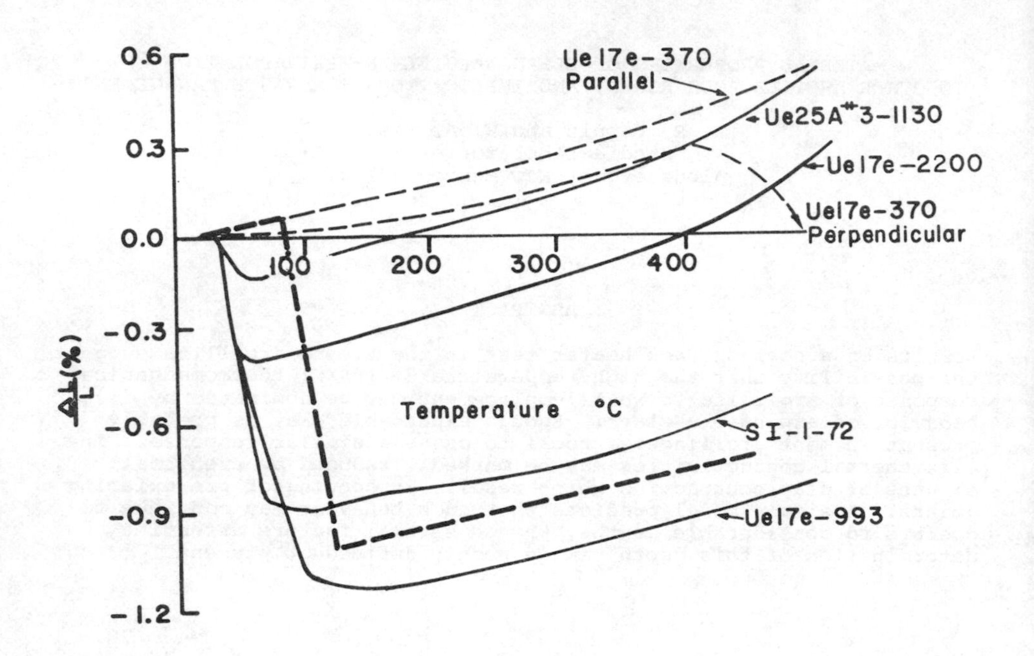

Figure 1. Ambient-pressure linear thermal expansion of Eleana argillite to 500°C. Solid lines = "natural state" samples, waxed and wrapped until prepared for measurement; Dashed lines = air-dried samples; Heavy dashed line = expansion behavior assumed in modeling of Eleana heater test.

Introduction

Results of a full-scale (3.5 kW) near-surface heater experiment recently completed in argillite of the Eleana Formation on the U.S. Department of Energy Nevada Test Site indicate that the thermo-mechanical response of many argillaceous rocks to the emplacement of heat-producing nuclear wastes may be dominated by effects resulting from the contraction of expandable clays. This conclusion is applicable only if power densities are sufficient to cause boiling of water within the formation or significant rock dehydration at lower temperatures, and is based on a consistent correlation between lab-measured thermal expansion of argillite and field observations related to the Eleana heater test. These observations indicate: 1) opening of joints in the heater hole wall, as a result of volumetric contraction; 2) apparent decrease of in situ thermal conductivity of the argillite below expected values for temperatures above approximately 100°C; 3) marked increase in argillite permeability inside the 100°C isotherm, from the millidarcy to Darcy range; and 4) presence of a rubbled zone near the heater, encountered in posttest drillback and coring operations.

The objectives of this paper are to: 1) examine the amounts of expansion/contraction that might be expected to occur for a range of argillaceous rocks on dewatering of the expandable clays; 2) estimate some of the effects of this contraction on in situ thermal conductivity; and 3) gain some feel for the extent to which tensional behavior may occur as a result of experimentation or waste emplacement in argillaceous rocks at depth.

Thermal Expansion/Contraction of Argillaceous Rocks

Thermal expansion of silicate rocks is complex. It may be strongly affected by variables such as bulk confining pressure, which affect the state of the matrix, porosity, and microcracks present,[1] and also by the local fluid or water pressure at the time of heating. The temperatures at which the expandable clays within argillaceous rocks dehydrate and contract are strongly dependent upon fluid pressure. Volume and length changes associated with this dehydration depend upon the amount of expandable clay present.

Measurements of the thermal expansion of argillaceous rocks are extremely limited in number. Results of a series of ambient-pressure measurements on Eleana argillite are shown in Figure 1. Sample SI-1-72 is from the central heater hole of the full-scale heater test, 1 m above the heater center plane. "Natural state" samples, which were waxed and wrapped until prepared for measurement, are indicated by solid lines in Figure 1 and show variable linear contraction beginning soon after heating. This contraction reflects the dewatering and contraction of expandable clays, which begin at very low temperatures in these tests conducted at ambient pressure. After reaching maximum contraction at or near 100°C, the samples then expand continuously to 500°C, with a linear expansion coefficient (α_L) of between 8 and 15 x 10^{-6} $^{o}C^{-1}$.

Anisotropy of expansion varies markedly in the Eleana. Sample Uel7e-2200 is not measurably anisotropic; three mutually perpendicular blanks of this sample have the same linear expansion coefficient within ± 1.0 x 10^{-6} $^{o}C^{-1}$, the approximate limit of accuracy of these measurements. Much of the Eleana appears to be similarly massive, with macroscopically indistinguishable bedding. Macroscopically layered samples, however, generally demonstrate anisotropic expansion. For example, sample Uel7e-993 contracts 1.2 ± 0.2% parallel to layering upon long-term heating at 110°C, and 1.9 ± 0.2% perpendicular to layering. These values correspond to α_L values of -130 and -210 x 10^{-6} $^{o}C^{-1}$, respectively, between ambient and 110°C.

Sample state also plays a major role in results of expansion measurements of argillaceous rocks. For example, Ue17e-370, which is strongly layered and was air-dried for approximately six months before measurement, does not show contraction at or below 100°C, but does display marked anisotropy above 400°C, presumably due to structural breakdown of constituent clays. Above 400°C, the blank of sample 370 cut perpendicular to layering contracts, while that cut parallel to layering continues to expand. Also shown in Figure 1 is the expansion behavior assumed for purposes of mechanical modeling of the full-scale heater test in the Eleana Formation. Note that it was assumed that dehydration of the in situ argillite did not begin until 75°C.

Mineralogical analyses of the Eleana argillite show a broad variation in assemblage. The general assemblage is: quartz + "illite" ± kaolinite ± chamosite ± pyrophyllite ± chlorite ± vermiculite ± siderite ± calcite ± ferroan dolomite ± pyrite. Quartz content ranges approximately from 15 to 50 weight percent. From a thermal expansion point of view, the most important phase in argillite would appear to be the "illite", which contains appreciable expandable interlayers of vermiculite or montmorillonite. Bulk X-ray analysis of powdered samples indicates that the basal spacing of the "natural state illite" is 10.5 to 10.8 Å at ambient temperature, decreasing to 10.0 - 10.2 Å after heating to 400°C. Based on the interpretive curves of Jonas and Brown,[2] this behavior indicates that the "illite" contains up to about 10% interlayering of chlorite, and that 60 - 80% of the "illite" is indeed illite (nonexpandable), while the rest is an expandable phase (montmorillonite or vermiculite).

While there is generally less expandable clay in Paleozoic shales than in younger material, it would appear that enough should be present in most older rocks,[3,4] such as Eleana argillite (Mississippian age), to result in there being no net expansion upon heating to 100°C or the boiling point of water. In fact, there may be a net contraction of many rocks. Assuming that the volumetric expansion coefficient (α_V) of an average argillaceous rock at temperatures below boiling is $36 \times 10^{-6}\ {}^{o}C^{-1}$, and that the boiling point is 100°C, then the net volumetric expansion between ambient temperature and the beginning of boiling will be approximately 0.003 cm^3/cm^3. If the basal spacing of the expandable clays (montmorillonite or vermiculite) present in this rock as discrete phases or as interlayers with a nonexpanding phase is assumed to contract from 12.5 to 10 Å upon dehydration near 100°C, then only 1.5 volume percent of the rock need be expandable clay in order for the rock to experience no net expansion to that temperature. Higher content and larger initial basal spacing of the expandable phase(s) would lead to absolute contraction. Retention of partially expanded basal spacings to higher temperatures would reduce the amount of contraction.

Considering an "average" shale,[5] which contains approximately 70% clay mineral, only 2.5% of the total clay content must be expandable to result in no net expansion at 100°C under these assumptions. If "illite" makes up 25% of the total clay assemblage, a reasonable lower limit,[4] then 10% interlayering of expandable phase in the "illite" is sufficient. If "illite" comprises 80% of the clay assemblage, which is quite common in Lower Paleozoic shales, then only 3% interlayering is required.

It thus appears as if virtually any argillaceous rock, even a Paleozoic shale generally considered to contain "no" expandable clay, may be subject to the same sort of contraction behavior as seen in the Eleana near-surface heater test. This is based on the fact that the smallest amount of expandable-phase interlayering that can be detected by X-ray analysis is approximately 10%,[6,7] and that at least slight interlayering is generally reported in even deep-seated shales. For example, assuming that the clay minerals are randomly oriented, illitic shales of the Oligocene Frio Formation, encountered in deep wells along the Gulf Coast of the United States, should undergo 0.4 ± 0.1% linear contraction, based on their mineralogical composition and reported uncertainties in extent of interlayering of expanded phases.[8] Representative Pierre Shale containing approximately 75% total clay, of which 75% is mixed-layer illite containing 60% montmorillonite interlayers,[9] should undergo 2.2 ± 0.3% linear contraction after dehydration of expandable phases.

Effects of Contraction on In Situ Conductivity

Contraction of argillaceous rocks near the boiling point of water will have an appreciable effect on in situ thermal conductivities only if the contraction of heated domains is not compensated by general infilling from surrounding areas. In the Eleana heater test, contraction of clays within discrete joint-bounded blocks appears to have resulted in simple opening of pre-existing joints, with perhaps some additional breakage as well. In the following discussion, similar behavior is assumed to be realistic for a range of argillaceous rocks, and thermal effects resulting from the opening of pre-existing joints are briefly discussed. A simple one-dimensional analysis of changes in thermal conductivity due to the presence of planar joints perpendicular to the direction of heat flow is given.

Little information is available concerning the matrix conductivity of argillaceous rocks at or above 100°C. Some of the available information is summarized in Table I. As shown, the reported range of conductivities at elevated temperature is from 0.3 to 2.1 W/m°C.

If it is assumed that heat flow within a rock mass is perpendicular to planar joints, that the joints are small enough in aperture to not allow convection, and that thermal radiation is negligible, then the effective conductivity of the jointed mass can be approximated by use of the formula for heat flow perpendicular to the layers of a laminated structure:

$$K = \frac{1}{\frac{X_1}{K_1} + \frac{X_2}{K_2}} \qquad 1.$$

where K_1 and K_2 are rock matrix conductivity and saturated steam conductivity at 100°C, respectively, and X_1 and X_2 are the fractional lengths of rock matrix and joint per unit length of rock mass. Thermal conductivities calculated as a function of percent linear contraction and matrix conductivity at 100°C, by means of equation 1, are shown in Figure 2. The specific matrix conductivities indicated are 0.3 W/m°C for the Boom clay (Belgium) and 1.75 for Eleana argillite parallel to layering. Curves are also drawn for initial matrix conductivities (100°C) of 0.75, 1.1 and 2.5 W/m°C. The solid curves in Figure 2 indicate that absolute decreases in conductivity are greatest at constant percent joint opening (or linear contraction) for rocks with the highest matrix conductivities. As indicated by curves a and b in Figure 2, greater contraction is also required to produce a constant relative % decrease in conductivity, as matrix conductivity decreases.

The effects of even small amounts of contraction could be significant. For example, if it is assumed that shale from the Frio Formation had a matrix conductivity of 2.0 W/m°C at 100°C, the estimated amount of contraction of this shale could reduce this to near 1.5 W/m°C, a 25% reduction. In the case of Eleana argillite, 1% contraction parallel to layering could reduce in situ conductivity from 1.75 to about 1.05 W/m°C, a 40% reduction. Contraction perpendicular to layering in the Eleana could reduce the effective conductivity from 1.5 to 0.75 W/m°C, a 50% reduction. If contraction and joint opening are not considered, the ratio of conductivities

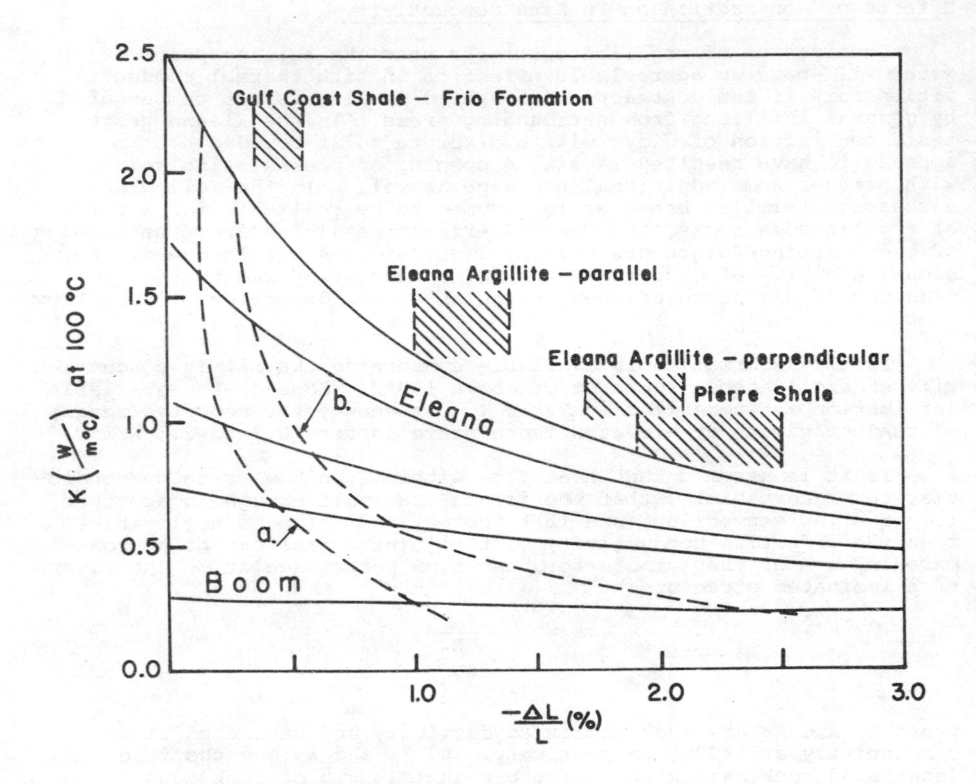

Figure 2. Approximate rock mass conductivity at 100°C as a function of matrix conductivity and extent of joint opening. Solid curves are drawn for matrix conductivities of 2.5, 1.75, 1.1, 0.75, and 0.3 W/m°C, and show absolute decreases as a function of contraction. Curves for conductivities of 1.75 and 0.3 W/m°C correspond to available data on Eleana argillite and Boom clay, respectively; other solid curves are for purposes of interpolation only. Dashed lines a and b indicate the amount of contraction required for constant 10% and 20% reduction of rock mass conductivity below matrix value, respectively. For example, approximately 0.4% contraction is required for 10% reduction in conductivity of a rock with matrix conductivity of 0.75 W/m°C. Indicated ranges of contraction of specific rocks are meant to imply nothing about their conductivity.

perpendicular and parallel to layering in the Eleana argillite (1.50/1.75) is 0.86 at 100°C. If contraction is considered and it is assumed that joints perpendicular to bedding are also perpendicular to heat flow in the layering, the ratio decreases to 0.71. Decreased sensitivity of in situ conductivity for rocks with a relatively low matrix conductivity to uncertainties caused by thermal contraction and jointing may partially offset advantages of waste emplacement in higher-conductivity media.

Additional factors would appear to add uncertainty to the estimates of in situ conductivity based solely on matrix conductivity and mineralogy. For example, heat flow in real or simulated waste emplacement, such as the Eleana heater test, is not constrained to be perpendicular to any joint sets, nor are joint sets necessarily continuous. Contraction of clays appears to result predominantly in opening of the pre-existing joint sets, rather than in generation of new joints. This fact obviously requires that future experimentation in argillaceous rocks include careful characterization of the pre-existing joint sets and consideration of the impacts of their possible effects on heat flow in different directions relative to canister orientation.

In the case of the Eleana argillite, data on fracture frequency and orientation taken from vertical drill holes indicate that some 70% of all fractures in the argillite are parallel to bedding.[15,16] At most depth intervals in these holes, however, there is a distinct subgroup of joints that is vertical or nearly so. Similar predominance of bedding plane joints appears to be the case in many argillaceous rocks.[17] Joint (bedding plane parting) frequencies on the order of 30/m are not uncommon parallel to layering. Steeply dipping joints, which tend to occur in two distinct sets, often occur at spacings of 1 m or more. Frequently, only one vertical set is well-developed, regular, and relatively continuous, and the other is much less prominent. One possible result of this is that considerable directional anisotropy may be superimposed on the thermal response parallel to layering in and above a repository in argillaceous rocks. For jointed rock masses as treated thus far, the calculated ratio of conductivity perpendicular and parallel to layering in the Eleana is 0.71. If, however, heat flow is parallel to the intersection of bedding and one vertical joint set, the ratio decreases to 0.43. This directional variation would cause the in situ temperature distribution to be a three-dimensional function of heat flow, clay content, and joint orientation.

Contraction and Tensile Joint Opening at Depth

While opening of joints due to volumetric contraction appears to have taken place in the Eleana heater test, the emplacement depths to which similar behavior would be observed in the Eleana, or any other argillaceous rock type, are unclear at present. A simple, one-dimensional analysis, in part compared with detailed thermomechanical modeling of the Eleana heater test, indicates that such opening is not likely to be solely a near-surface phenomenon. In this analysis, the maximum tensile stress resulting from heating is taken to equal the product of the Young's modulus and maximum linear contraction of a given rock type, and is compared with assumed hydrostatic pressures as a function of depth. It is further assumed that tensional joint opening will occur if the tensional stress exceeds the sum of the hydrostatic stress plus in situ tensile strength of the rock. The resulting failure criterion, for an assumed in situ tensile strength of 3.5 MPa (500 psi), is shown in Figure 3 and compared with the results of thermomechanical modeling of a configuration the same as the Eleana heater test, at a series of depths. In this figure, the full and open circles represent results of two-dimensional axisymmetric thermomechanical modeling of the heater test.[18] The calculations were made with the ADINA code,[19] which includes a tension cut-off. For these calculations this was set at 3.5 MPa.

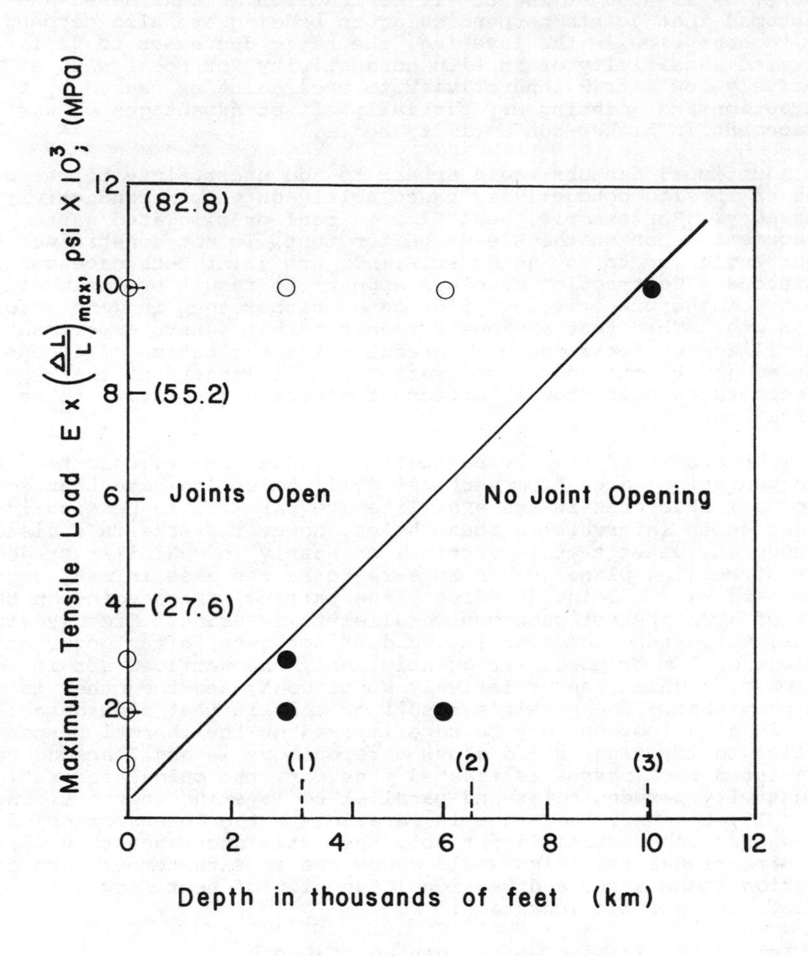

Figure 3. Maximum possible thermally induced tensile stress versus depth. Above solid line, joints are predicted to open, due to volumetric contraction. Below it they are not.

The simplified assumptions predict that, for an assumed Young's modulus and maximum linear contraction of the Eleana of 6.90 GPa (1×10^6 psi) and 1.0 percent, respectively, the maximum possible tensile stress should be 69 MPa (1×10^4 psi), and some tensional opening should occur for emplacement at depths up to 2900 m (9500 feet). Variations in clay content will change the depths to which opening persists for a rock of constant Young's modulus. For example, a rock which has a Young's modulus of 6.90 GPa, but which contracts only 0.1% upon dehydration, should exhibit a maximum tensile stress of only 6.90 MPa (1×10^3 psi), and opening of joints to a depth of only about 500 m. Calculation of variation in the depth to which tensile openings persist as a function of clay content are relatively straightforward.

Many factors, however, will influence the in situ Young's modulus of argillaceous rock and, in turn, cause changes in the depth boundary between open and closed joints shown in Figure 3. Intact samples of most shales have unconfined Young's moduli in the range of 0-50 GPa.[20] The lab-measured modulus of the Eleana, 3.7 GPa, falls near the low end of this range and is also low compared to a "typical illitic shale"; see Table II. However, the modulus of the Eleana increases rapidly with increasing confining pressure, and for pressures greater than about 20 MPa, is constant at about 16 MPa. The phenomenon of a constant modulus above a certain confining pressure has also been observed in other argillaceous rocks.[21,22] It should be pointed out that the depth corresponding to the pressure at which the modulus of the matrix becomes constant ($\sim$ 0.9 km) is typical of depths discussed for potential repositories and, hence, that a pressure-dependent modulus may be operative at the disposal horizon. A pressure-dependent modulus obviously should be used in mechanical modeling of overlying layers.

Moisture content has recently been shown to be a very important factor in altering the elastic compliance matrices for three argillaceous rocks from coal mines in the U.S.[21] The tests were run in unconfined compression at room temperature. When the relative humidity was varied from 0 to 100%, Young's modulus measured perpendicular to bedding was reduced by at least 50% for each of the three shales. Some of the compliances were even more strongly affected. The implication of this sensitivity is that the moduli of argillaceous rock can be time-varying due to changing water contents, even at a given P and T.

In situ stress, both pre-existing and resulting from repository excavation and waste emplacement, may have a strong influence on thermal conductivity of argillaceous rocks, through its effect on dilatancy (microcrack opening) and joint dilatancy. Microcrack dilatancy depends on both the orientation and magnitude of the deviatoric part of the stress tensor (e.g., reference 24). Joint dilatancy, on the other hand, results from the riding of adjacent blocks over asperities on the joint surface between them, and is more closely related to the amount of shear displacement.[25]

Our findings to date with regard to stress-volume relationships for Eleana argillite are summarized in Figure 4. The specimens tested were right circular cylinders 5 cm in diameter x 10 cm long and, though coherent, contained macroscopic joints at a natural joint spacing of about 2.5 cm. Several distinct joint orientations were evident in each sample, giving the specimens a blocky appearance. In Figure 4, the vertical axis is mean stress ($\sigma_m = (\sigma_1 + \sigma_2 + \sigma_3)/3$) and the horizontal axis is volumetric strain, ε_V, in percent. Compaction is considered to be positive. The thick line is the common hydrostat for the four tests shown. The tests were run in confined compression, so that the first phase of each test is hydrostatic loading up to the predetermined confining pressure (0, 20, and 38 MPa), marked by triangles in the figure. The thin lines, one for each test, show the $\sigma_m - \varepsilon_V$ relations for the deviatoric stress

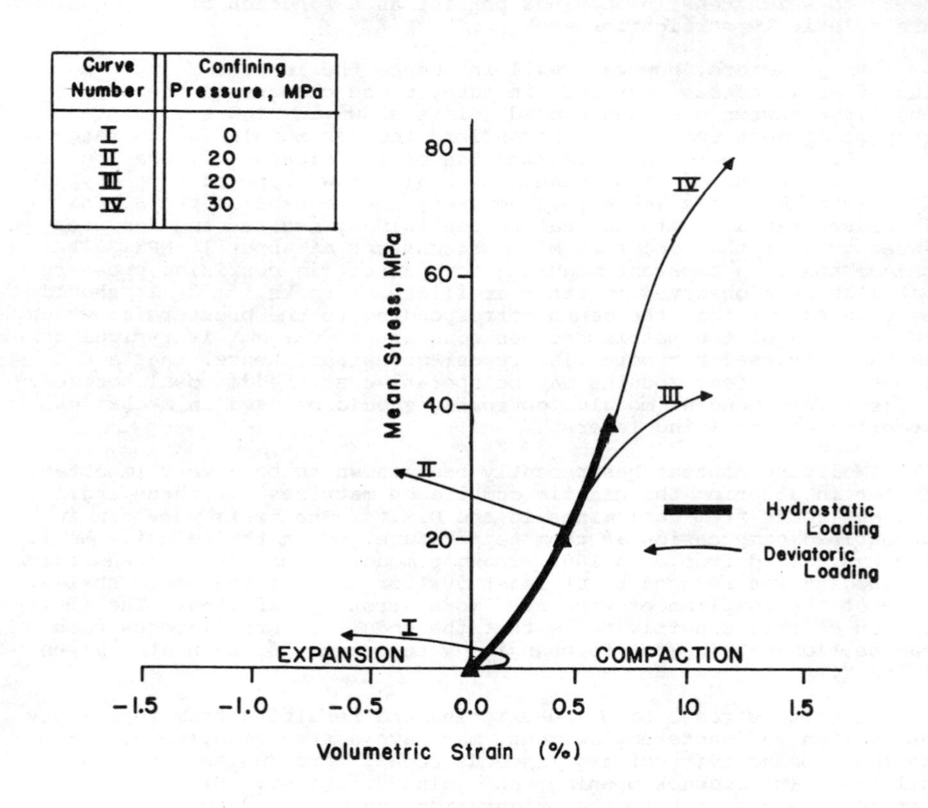

Figure 4. Mean stress versus volume for ambient-temperature testing of Eleana argillite. Compaction is considered to be positive.

phase of each test. Positive slopes represent compaction, and negative slopes represent expansion or dilatancy. Curve I represents results of an unconfined compression test and shows that initial axial compression causes a decrease in volume, as a result of crack closure. At about 50% of the specimen strength, dilatancy sets in and the volume increases continuously until the ultimate strength is reached. Test IV, which was run at a confining pressure of 38 MPa, shows that the specimen compacts during hydrostatic loading, and continues to compact during deviatoric loading. Transitional behavior is exhibited by tests II and III, run at a confining pressure of 20 MPa. Test II showed slight compaction during early deviatoric loading, followed by expansion; in contrast Specimen III showed continuous compaction throughout. Apparently 20 MPa (∿ 3000 psi) represents a critical confining pressure above which specimens of Eleana argillite compact in response to deviatoric loading at room temperature, and below which they dilate.

If a repository in Eleana argillite were less than 900 m (3000 feet) deep confining pressures would be less than ∿ 20 MPa, and dilatancy could be a predominant response to deviatoric stresses. Also, near-field stresses resulting from excavation and waste emplacement should tend to be more deviatoric than those of the far-field. Hence, dilatancy problems should be enhanced in the near-field, i.e., in that area that is most likely to undergo sufficient heating to cause contraction of expandable clays. The effect of both such heating and deviatoric loading will be to enhance cracking, with consequent decrease in thermal conductivity and increase in permeability.

At least two major factors that might affect the response of argillaceous rock to waste emplacement remain poorly defined. Little work has been done on the mechanical properties of argillaceous rock at elevated temperature. Although it has been shown[22] that the strength of some shales decreases markedly with increased temperature, the temperature dependence of the modulus is not documented. For many other rock types, in fact, increasing confining pressure seems to have a stronger effect on elastic moduli than increasing temperature. The effect of time, as examined in creep tests or in constant deformation rate tests of argillaceous rocks, is also not well documented. One recent study[25] does provide a limited amount of creep data on the Ophir Shale. Although some of the results of this study are difficult to interpret, they do indicate that steady-state creep rates at very low deviatoric stresses are on the order of 10^{-9} s^{-1}. One observation of this study is that, when the shale is subjected to a hydrostatic pressure step-function history, the resulting volume-time curve looks just like an ordinary creep curve, i.e., volume compaction has a time-dependent, viscoelastic nature. This is an important result because most creep testing programs have ignored the fact that the general viscoelastic constitutive relation of rocks includes a time-dependent bulk modulus, as well as the time-dependent Young's modulus.

Conclusions

Though the treatment present here is simple and subject to many uncertainties and uncontrolled or poorly defined variables, three main conclusions appear to be justified by available data:

1) Expandable clays, such as montmorillonite or vermiculite, may play a strong role in the thermomechanical response of argillaceous rocks to the emplacement of nuclear wastes. This is especially true if the rocks are heated above the boiling point of water. Expandable phases may occur as discrete phases, or as interlayers in mixed-layer clays, even at small degrees of interlayering commonly not detectable by means of X-ray diffraction.

2) Dewatering and contraction of expandable clays, if accommodated by opening of the pre-existing joints inherent to argillaceous rocks, could cause a reduction of in situ thermal conductivity. Both absolute and relative decreases would be greater for rocks with higher matrix conductivity. Variations in directional heat flow relative to pre-existing joints could result in a highly anisotropic thermal distribution. Decreases of in situ conductivity as a result of heating due to waste emplacement could occur for disposal at considerable depths. At constant Young's modulus, the depth to which joint opening would continue to occur decreases with decreasing content of expandable clay.

3) In addition to the effects of variable Young's modulus and clay content, variables such as confining pressure, non-hydrostatic stress, rock moisture content, time, temperature, and deformation rate will influence the presence or absence of dilatant behavior at depth. These factors are poorly defined at present.

It should be kept in mind that this paper only addresses uncertainties in the utilization of argillaceous rocks as emplacement media for nuclear wastes. There is nothing inherent in these concerns that would appear at present to eliminate argillaceous rocks from further consideration. However, these uncertainties must be resolved before a repository in argillaceous rock could be designed with confidence.

Table I

Matrix Thermal Conductivity of Argillaceous Rocks at or Above 100°C

Rock Types	K (W/m°C)	Reference
Boom Clay (Belgium)	∿ 0.3	10
Slate (Perpendicular to bedding)	1.76	11
Eleana Argillite		
(Parallel to Layering)	∿ 1.80	12
(Perpendicular to Layering)	∿ 1.48	
Conasauga Shale (calcareous)	0.7-2.1	13,14

Table II

Comparison of Eleana Argillite to "Typical Illite Shale;" Shale Data from Reference 17

	Eleana	Illite Shale
Unconfined		
Young's Modulus (GPa)	3.74	11.03
Poisson's Ratio	0.30	0.15
Compressive Strength (MPa)	4.75	69.0
Cohesion (MPa) (Intersection of Mohr-Coulomb Failure Envelope with Shear Stress Axis.)	0.8	41.0
Confined		
Bulk Modulus (GPa)	14.0	5.20
Angle of Internal Friction (deg)	35	26

References:

1. Cooper, H. W., and Simmons, G.: "The Effects of Cracks on Thermal Expansion of Rocks," Earth and Planetary Science Letters 36, 404-412 (1977).

2. Jonas, E. C., and Brown, T. E.: "Analysis of Interlayer Mixtures of Three Clay Mineral Types by X-Ray Diffraction," Journal of Sedimentary Petrology 29, 77-86 (1959).

3. de Segonzac, G. D.: "The Transformation of Clay Minerals During Diagenesis and Low-Grade Metamorphisis: A Review," Sedimentology 15, 281-346 (1970).

4. Weaver, C. E.: "The Significance of Clay Minerals in Sediments," Fundamental Aspects of Petroleum Geochemistry, B. Nagy and U. Columbo, Eds., pp. 37-75, Elsevier (1967).

5. Shaw, D. B., and Weaver, C. E.: "The Mineralogical Composition of Shales," Journal of Sedimentary Petrology 35, 213-222 (1965).

6. Weaver, C. E.: "Distribution and Identification of Mixed-Layer Clays in Sedimentary Rocks," American Mineralogist 41, 202-221 (1956).

7. Hower, J. and Mowatt, T. C.: "The Mineralogy of Illites and Mixed-Layer Illite/Montmorillonites," American Mineralogist 51, 825-854 (1966).

8. Hower, J., Eslinger, E. V., Hower, M. E., and Perry, E. A.: "Mechanism of Burial Metamorphism of Argillaceous Sediment: 1. Mineralogical and Chemical Evidence," Geological Society of America Bulletin 87, 725-737 (1976).

9. Schultz, L. G.: "Mixed-Layer Clay in the Pierre Shale and Equivalent Rocks, Northern Great Plains Region," United States Geological Survey Professional Paper 1064-A, 28 p. (1978).

10. "Belgian Working Draft on a Repository of Solidified Nuclear Waste in a Deep Tertiary Clay Formation," Studiecentrum Voor Kernenergie, Mol, Belgium, WG.7/45 (1979).

11. Clark, S. P., Jr. (Ed.): Handbook of Physical Constants, Geological Society of America Memoir 97, 587 p. (1966).

12. McVey, D. F., Thomas, R. K., and Lappin, A. R.: "Small-Scale Heater Tests in Argillite of the Eleana Formation at the Nevada Test Site," SAND 79-0344, Sandia Laboratories, Albuquerque, NM (in press).

13. Smith, D. D.: "Thermophysical Properties of Conasauga Shale," Y-2161, Oak Ridge National Laboratory, Oak Ridge, Tennessee (1978).

14. Krumhansl, J. L.: "Preliminary Results Report--Conasauga Near-Surface Heater Experiment," SAND 79-0745, Sandia Laboratories, Albuquerque, NM (1979).

15. Murphy, B. J., Chormack, M., and Hoover, D. L.: "Engineering Geology of the Eleana Heating Experiment Site, Area 17, Nevada Test Site," United States Geological Survey Open File Report USGS-474-265 (in press).

16. Hodson, J. N., and Hoover, D. L.: "Geology of the UE17e Drill Hole, Area 17, Nevada Test Site," United States Geological Survey Open File Report USGS-1543-2 (1979).

References:

17. "Technical Support for GEIS: Radioactive Waste Isolation in Geologic Formations, Volume 6, Baseline Rock Properties--Shales," Office of Waste Isolation, Oak Ridge, Tennessee, Y/OWI/TM-36/6 (1978).

18. Thomas, R. K., Sandia Laboratories, Albuquerque, NM, Personal Communication (1979).

19. Bathe, K. J.: "ADINA, a Finite Element Program for Automatic Dynamic Incremental Nonlinear Analysis," Massachusetts Institute of Technology Report 82448-1 (1977).

20. Lama, R. D., and Vutukuri, U. S.: Handbook on Mechanical Properties of Rocks 2, 436-441, Tech. Publications, Germany (1978).

21. Bredthower, R. O.: "Strength Characteristics of Rock Samples under Hydrostatic Pressure," Transactions of American Society of Mechanical Engineers 79, 695-708 (1957).

22. Handin, J., and Hager, R. V., Jr.: "Experimental Deformation of Sedimentary Rocks under Confining Pressures: Tests at Room Temperature on Dry Samples," Geological Society of America Bulletin 41, 1-50 (1957).

23. Van Eeckhaut, E. M., and Peng, S. S.: "The Effects of Humidity on the Compliance of Coal Mine Shales," International Journal of Rock Mechanics and Mining Sciences and Geomechanics Abstracts 12, 335-340 (1975).

24. Brace, W. F.: "Micromechanics of Rock Systems," in Te'en, M. (Ed.), Structure, Solid Mechanics, and Engineering Design, Wiley, New York, 187-204 (1971).

25. Goodman, R. E., and Dubois, J.: "Duplication of Dilatancy in Analysis of Jointed Rocks," Journal of the Soil Mechanics and Foundations Division, American Society of Civil Engineers 98, 399-422 (1972).

26. Cogan, J.: "Triaxial Creep Tests of Opohonga Limestone and Ophir Shale," International Journal of Rock Mechanics and Mining Sciences and Geomechanics Abstracts 13, 1-10 (1976).

Discussion

R. PUSCH, Sweden

Do you have any idea of what K_o (ratio of horizontal and vertical stresses in situ) might exist at greater depth in sediments containing expanding clay minerals ? Increased temperature should lead to a reduced viscosity and thus to the opposite effect you mention especially since I believe K_o would be higher than you said. In fact, I think such "healing" would be one of the few advantages in placing repositories in natural sediments.

A.R. LAPPIN, United States

Certainly, the conventional wisdom concerning argillaceous rocks in that they cannot support deviatoric stresses at depth, and as I understand it, this has generally been taken as an implicit assumption. Intuitively, increased temperature would be expected to decrease the ability to sustain stresses. My intention here, however, has been to look at some of the consequences of such healing not taking place, since I am certain we did not see it in our near-surface test. The consequences of that fact suggest a basic phenomenology that had not been considered before the test. Certainly, the only way we will fully understand the behavior of these rocks at depth is by means of testing at depth.

A.G. DUNCAN, United Kingdom

Through what distance would you expect the tensile load to be transmitted in these argillaceous rocks ?

A.R. LAPPIN, United States

In the numerical modeling of the test, the tensional zone included only the volume inside the 80-100°C isotherm. Incidentally, this is consistent with results of far-field modeling of a generalized repository in argillite ; while there was volumetric contraction within the 100°C isotherm, this zone was surrounded by a continuous zone of compressive stresses.

A.G. DUNCAN, United Kingdom

Does this not indicate that we should not be too concerned about what happens to the rocks in the immediate vicinity of the heat source ?

A.R. LAPPIN, United States

In some respects, yes. For example, if it can be shown that the presence of the contracting zone in no way affects the ultimate containment capability of the rock, then the presence of a fractured, highly permeable region may be of no concern. In fact, since many of the geochemical studies appear to indicate that the amount of nuclide sorption is proportional to surface area, it may actually help, since the effective surface area would certainly be increased within the zone where the rock had contracted. On the other hand, our approach and, as I understand it, the overall DOE approach, has been heavily weighted towards the feeling that we must ultimately be able to model the rock or repository response in great detail. If one takes this as an actual requirement then we must understand the near-field responses very well, regardless of whether or not they seriously affect the ultimate containment.

F. GERA, NEA

Do you expect the changes in the rock to be reversible ?

A.R. LAPPIN, United States

Well, as mentioned previously, the swelling clays apparently retain their ability to rehydrate after being heated to temperatures in the order of 200-250°C. To my knowledge, however, this conclusion is based solely on rather short-term heating experiments. Whether or not the clays would rehydrate after very long times at elevated temperature remains to be seen.

TEST RESULTS AND SUPPORTING ANALYSIS
OF A NEAR SURFACE HEATER EXPERIMENT
IN THE ELEANA ARGILLITE

By
D. F. McVey, A. R. Lappin and R. K. Thomas
Sandia Laboratories
Albuquerque, New Mexico

ABSTRACT

A preliminary evaluation of the in situ thermomechanical response of argillite to heating was obtained from a near-surface heater test in the Eleana Formation, at the United States Department of Energy, Nevada Test Site. The experiment consisted of a 3.8 kW, 3-m long x 0.3-m diameter electrical heater in a central hole surrounded by peripheral holes containing instrumentation to measure temperature, gas pressures, and vertical displacement. A thermal model of the experiment agreed well with experimental results; a comparison of measured and predicted temperatures indicates that some nonmodeled vertical transport of water and water vapor occurred near the heater, especially at early times. A mechanical model indicated that contraction of expandable clays in the argillite produced a region 1.5 - 2.0 m in radius, in which opening of preexisting joints occurred as a result of volumetric contraction. Results of thermal and mechanical modeling, laboratory property measurements, experimental temperature measurements, and post-test observations are all self-consistent and provide preliminary information on the in situ response of argillaceous rocks to the emplacement of heat-producing nuclear waste.

*This work was supported by the U. S. Department of Energy (DOE) under Contract DE-AC04-76DP00789.

A U. S. DOE Facility

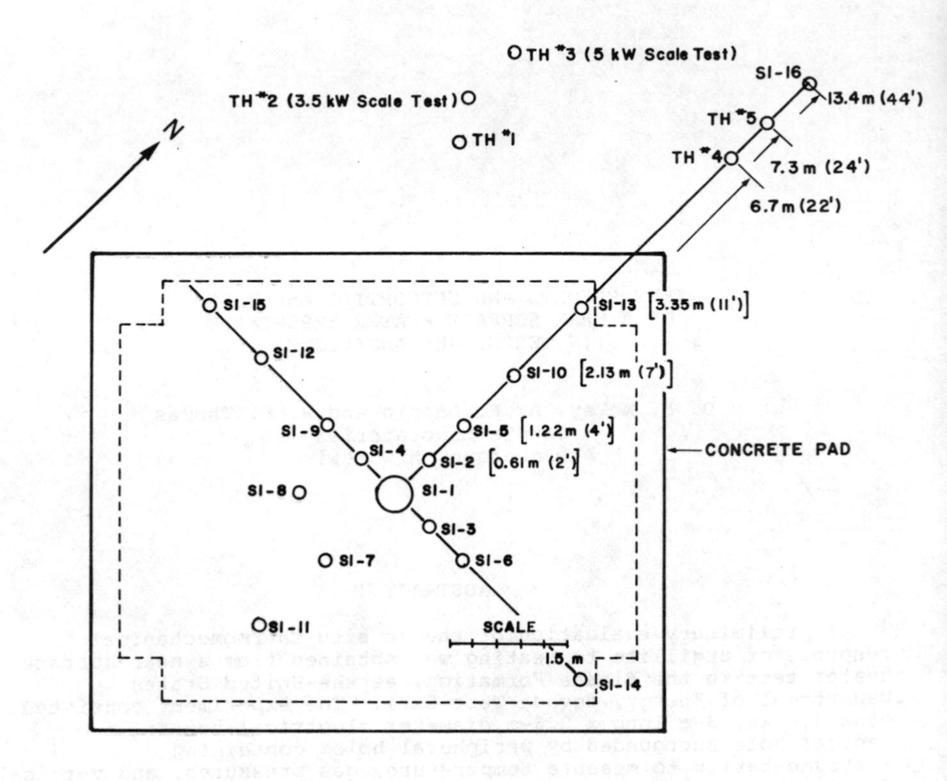

FIGURE 1: PLAN VIEW OF ELEANA NEAR-SURFACE HEATER SITE, NEVADA TEST SITE

Introduction

The United States Department of Energy is sponsoring a project at the Nevada Test Site (NTS) [1] to investigate the suitability of several "hardrock" media for use as a high-level waste repository. One medium being considered is argillite, a slightly metamorphosed shale which occurs within the Eleana Formation. Argillite is attractive as a host geologic medium for a waste repository because of its relatively high ion sorption characteristic [2], low matrix and joint permeability [3], reasonably high thermal conductivity and moderately high strength [4]. Specific argillite occurrences include Syncline Ridge in the north central region of NTS and the Calico Hills in the southwestern region of NTS.

The investigations at Syncline Ridge included installation and operation of an electrically powered near-surface heater test. A near-surface test was used in the absence of a suitable deep underground facility as the fastest and most economical method of gaining preliminary information on field response of argillaceous rocks. The objectives of the experiment were: 1) to obtain basic phenomenological data needed for the preliminary evaluation of the waste storage potential of argillaceous rocks; 2) to assess the predictive capability of available codes for analysis of the thermomechanical response of geologic media to waste emplacement; 3) to evaluate the applicability and reliability of laboratory-measured properties; and 4) to assess experimental techniques to aid in defining future needs.

This paper briefly discusses the test setup, some of the experimental results, and the supporting thermal and mechanical analyses for the Eleana full-scale heater test. Phenomena thought to influence the agreement between computations and experimental results are discussed along with the applicability of the laboratory property data to the in situ test.

Test Setup and Operation

A schematic of the Eleana heater site is shown in Figure 1. The heater, 0.32-m diameter x 3.0-m heated length, was installed in a 24.4-m deep x 0.356-m diameter hole, Sl-1, drilled in the argillite Sl-1 was extended, at 0.20-m diameter, to a total depth of 30.5 m. The heater dimensions roughly approximate those of waste canisters currently being considered. The heater hole was cased to the bottom of the obviously weathered zone (approximately 16 m deep) with steel pipe set in grout.

The instrumentation holes Sl-2 through Sl-15 are 0.1 m in diameter. Holes TH-1 through TH-5 are test holes. TH-2 and TH-3 were used for two small scale heater experiments run prior to the full-scale experiment. The instrumentation ray containing holes Sl-11 and Sl-16 is parallel to strike and the ray containing Sl-15 and Sl-14 is normal to strike. The formation dips about 30° to the southeast. The principal instrumentation plane is parallel to strike. Instrumentation installed within the holes consisted of thermocouples for temperature measurement, vertical extensometers in holes Sl-7, Sl-9 and Sl-11, and air pressure gauges at the top of holes Sl-1, Sl-3, Sl-6, Sl-7, Sl-9, Sl-11, and Sl-14. Of the experimental data obtained, only the temperatures will be discussed in detail in this report.

Thermocouple arrays were installed parallel (35 thermocouples) and perpendicular (20 thermocouples) to strike. The Type E (Chromel/Constantan) thermocouples were sheathed with 321 stainless steel. To install the thermocouples in the temperature instrumentation holes, the sheaths were strapped to the outside of 0.03-m o.d. plastic pipe, so that when the pipe was inserted to the

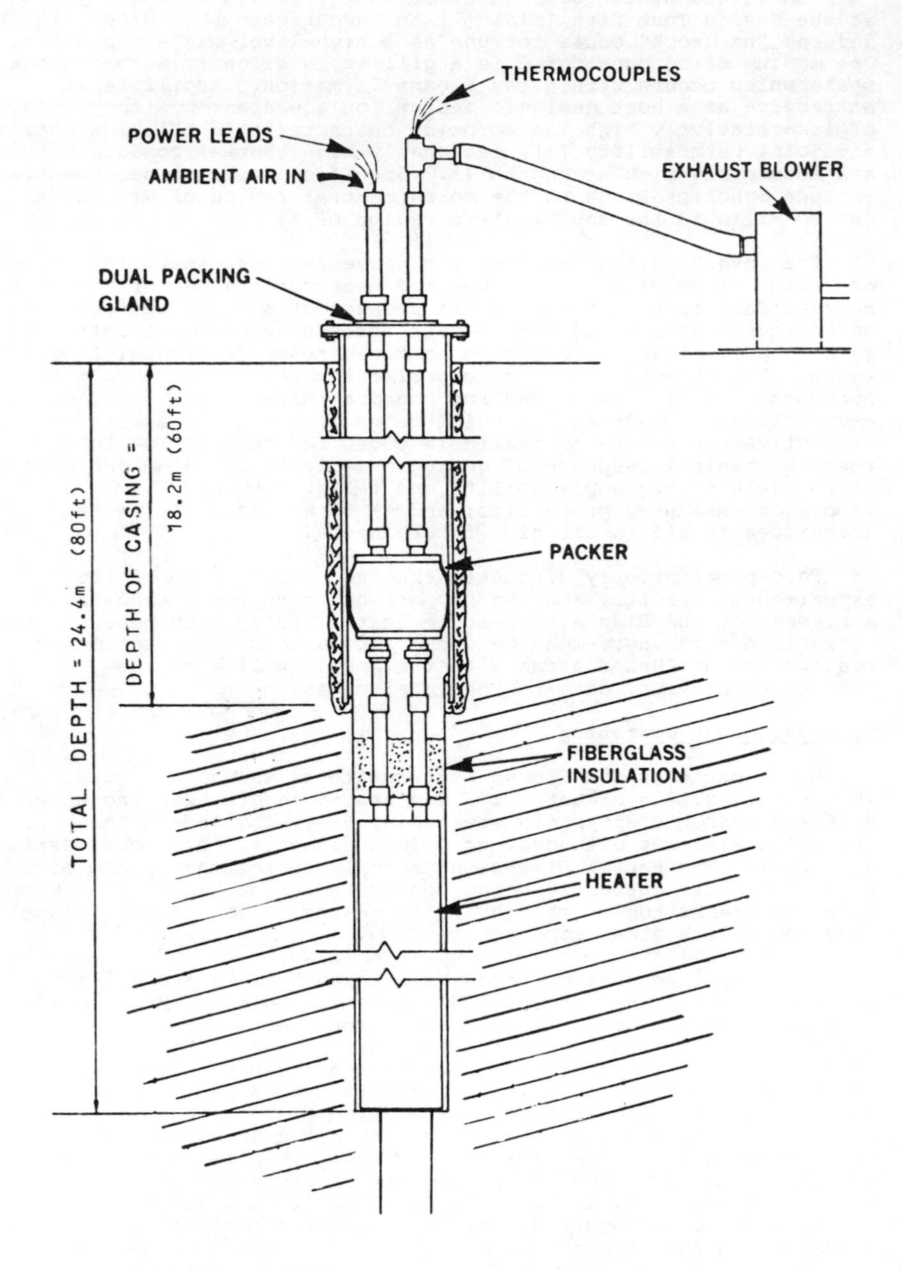

FIGURE 2: CROSS SECTION OF COMPLETE HEATER ASSEMBLY, EMPLACED

bottom of the hole, the junction was located at the correct depth. A grout with thermal properties which closely matched those of the argillite was then pumped down the pipe and allowed to fill the hole. Displacement and pressure measurement holes were not grouted.

A cross section of the heater assembly after emplacement in hole S1-1 is shown in Figure 2. The heater rested on the bottom shoulder of the 0.36-m diameter hole. Fiberglass insulation was inserted to about 0.3 m above the top of the heater to prevent convective currents from transporting heat to the packer used as a pressure seal at the bottom of the casing.

The heater was a sealed stainless steel (SS 304) pipe 3.8 m long. The top 0.2 m, containing the power terminals, was air cooled through the pipes containing the power and thermocouple leads. The next 0.6-m section contained vermiculite insulation and non-heated portions of the heater elements. The bottom 3.0 m was heated radiatively by three Chromalux tubular-hairpin elements, each rated at 6 kW at 815°C. A complete backup set of elements was contained in the heater, but was never required. Type E thermocouples were placed at several locations on the Chromalux heating elements, at several points within the junction and cold section compartments, and externally on the skin of the heater, at 0.09, 0.76, 1.5, 2.3 and 2.96 m from the bottom. The 1.5-m surface location was instrumented with three thermocouples. An additional two spring-loaded thermocouples, strapped externally to the heater at 1.5 m from the bottom, were thermally released at a heater skin temperature of about 95°C to spring against the hole wall and measure the rock surface temperature at the centerplane of the heater.

A constant electrical power of 3.5 kW was desired for the test. This power approximates the initial heat production from a similar size canister filled with ten-year-old high-level waste encapsulated in borosilicate glass, with 30% fission product loading. A test time of about one year was selected to allow the rock temperature to approach the maximum anticipated temperature.

Test Analysis

The analysis of the Eleana heater test covered several phases: conceptual design, detailed design, response prediction, final data correlation and model extension. At each phase, more and better property data were available, the interpretation and understanding of the data improved, and the calculational models were extended. This section will describe the latest thermal and mechanical models, discuss the laboratory derived properties used in the analysis and compare the computed results with the test data.

Thermal Model

COYOTE, a non-linear, two-dimensional, (planar or axisymmetric) finite element, thermal conduction, computer program [5] was used as the thermal analysis tool for the heater experiment. An axisymmetric geometry was assumed for modeling the Eleana heater test. The finite element configuration consisted of 360 quadrilateral elements, each with eight nodal points, four at the corners and one on each side. The analysis region was a cylinder of 13.5-m radius and 42.6-m length, with the 0.178-m radius by 22.8-m deep heater hole in the center. The outer dimensions were large enough so that the boundary could be treated as adiabatic for the duration of the test. The initial temperature of the region was set at 18°C, the ambient temperature at the experiment depth. The argillite was assumed to be uniform and orthotropic in the analysis region; e.g., weathered zones, non-uniformities and dip

TABLE I

THERMAL CONDUCTIVITY OF ELEANA ARGILLITE
(W/m°C)

T(°C)	A	B	C	D	E
25	-	-	2.43	-	-
50	-	-	2.17	2.3	1.90
75	1.79	1.68	-	-	-
100	1.78→1.69*	1.67→1.54*	2.06	2.2	1.70
150	1.53	1.44	1.80	1.73	1.53
200	1.45	1.39	1.67	1.73	-
250	1.39	1.32	1.69	1.73	-
300	1.36	1.28	1.73	1.73	-
350	1.33	1.24	1.61	1.73	1.33
400	1.31	1.20	1.46	1.60	-
450	1.29	1.17	1.48	1.50	1.29
500	-	-	1.27	1.40	-

A. Axial thermal conductivity, sample UE17e364.

B. Axial thermal conductivity, sample UE17e372.

C. Average of <u>radial</u> conductivities of sample UE17g80, UE17e623, UE17e627, and UE17g68.

D. Radial values used in analysis.

E. Axial values used in analysis.

*Decrease in conductivity, indicated by arrow, resulted from holding samples near 100°C for 24 hours.

TABLE II

SPECIFIC HEAT OF ELEANA ARGILLITE TO 500°C

T (°C)	Cp (cal/gm°C)	Used in Analysis** Cp (cal/gm°C)
75	0.23-0.33	0.28
100	0.25-0.40*	0.27
150	0.20-0.24	0.25
400	0.17-0.23	0.20
420	(-)0.56-(+)0.25***	0.20
475	0.18-0.25	0.20
510	0.23-0.39	0.20

*Includes effect of vaporization of an unknown amount of pore water. Smoothed out for this analysis.

**0.290 cal/gm°C added to Cp between 50 and 105°C to account for vaporization of water.

***Data scatter at this temperature due to exothermic sulfide oxidation.

were neglected. Since COYOTE only treats conduction in the solid rock, heat transfer due to motion of liquid or vapor within the rock was necessarily ignored.

Heat transfer across the air gap between the heater and the rock was assumed to be via radial one-dimensional radiation and conduction. Convection was ignored because the "effective conductivity" (ratio of the total heat transferred across a gap by convection and conduction to that transferred by conduction alone) is unity for the range of gap Rayleigh numbers encountered in the experimental configuration [6]. The heater emissivity and air thermal conductivity were specified as functions of temperature. Heater emissivity was taken from data [7] for 321 stainless steel oxidized in air at red heat for 30 minutes. The emissivity ranged from 0.205 at 300°K to 0.370 at 900°K. Total normal emissivity of the argillite surface was assumed to be 0.9. The mean argillite density was 2.63 gm/cc [8].

Conductivity and specific heat measured on Eleana Argillite samples are given in Tables I and II, respectively [8]. The smoothed conductivity data in columns "D" and "E" of Table I were used in the analysis. The conductivity at intermediate temperatures was obtained by linear interpolation between points.

It is of interest to discuss the applicability of the existing thermal property data base to the present application vis-a-vis an actual repository design. All thermal conductivity data available were obtained at ambient pressure, which has two important implications for utilization of the data. First, significant dehydration occurred in the sample during preparation and testing. Dehydration is reflected in decreasing radial conductivities between 25 and 100°C. In contrast, a waste repository might be below the water table thus inhibiting dehydration and pore water loss. Secondly, the portion of the conductivity decrease due to the presence of microcracks, which become filled with air or vapor as the sample dehydrates, would probably not occur at depth since the microcracks would be closed by the lithostatic pressure. Bulk compressibility data on Eleana argillite indicate that microcracks are essentially closed at 10.3 MPa confining pressure (about 430 m depth), suggesting that the conductivity decrease would not be as significant at depth. The laboratory conductivity data are, however, applicable to the near-surface test.

In addition, it must be noted that conductivities in Table I are for measurements parallel and radial to the long axis of available core material, rather than parallel and perpendicular to bedding. In both samples, bedding was inclined at approximately 30° to the sample base. Thus, the radial conductivity measurements, made by means of a transient line source technique, reflect some heat transport across layering, and are probably slightly low for transport parallel to layering. Likewise, the axial conductivity measurements, made by means of a guarded end plate technique, are probably slightly higher than for transport truly perpendicular to bedding. The error due to measuring the conductivity at roughly 30° to the material principle axes is expected to be less than 10%.

Specific heat data (Table II) were modified for use in the model by removing the singular spike at 100°C and replacing it with the heat of vaporization of 3% by weight water, added to the specific heat of the dry matrix. This mass averaging was required because the technique used in obtaining the data in Table II does not account for water evaporation during sample preparation. For compatibility with the computer code, vaporization of the water within the argillite was assumed to occur over the temperature range of 55°C to 105°C.

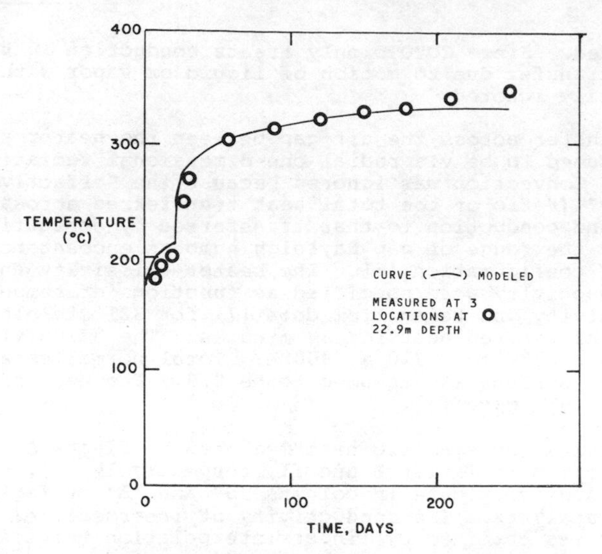

FIGURE 3: COMPARISON OF CALCULATED AND MEASURED HEATER SKIN TEMPERATURES AT HEATER CENTER PLANE

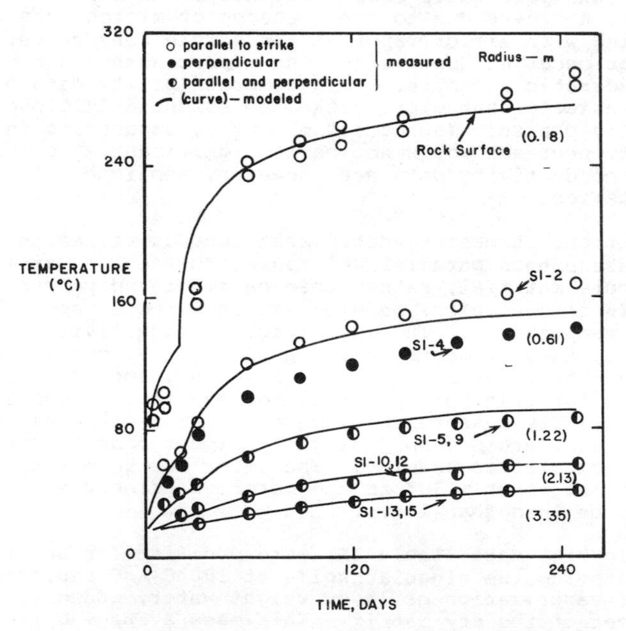

FIGURE 4: COMPARISON OF CALCULATED AND MEASURED ROCK TEMPERATURES AT THE HEATER CENTER PLANE, PARALLEL AND PERPENDICULAR TO STRIKE

Thermal Results

The COYOTE code was run with the previously described model, property data and the power input history of the test. The first 21 days of the test were run at 2.5 kW due to a misinterpretation of a power meter reading. After 21 days, the total power was increased to 3.8 kW. Based on cooling air temperatures, we assumed 3.5 kW was going into the rock, and 0.3 kW was being picked up by the terminal cooling air.

Temperatures as a function of time were computed at each nodal point of the COYOTE finite element grid The code output included temperature history plots for selected nodes, and isotherm contour-plots at selected times. Figure 3 compares computed and measured heater surface temperatures at the the heater midplane as a function of time; Figure 4 compares computed and measured temperatures at the rock surface and four locations radially outward from the heater midplane as a function of time; and Figure 5 compares isotherms parallel to strike at 30, 60, 100, 150, 200, and 250 days. The experimental isotherms were determined by interpolation of curves smoothed between the data points whereas the calculated results were obtained directly from the computer output. The large change in slope noted in Figures 3 and 4 at 21 days results from the power increase at that time.

Several general observations are derived from Figures 3, 4, and 5:

1) The character of the predictions and data match well considering the material variability, the small thermal property data base, and the fact that the geometry of the experiment is really three-dimensional (due to the inclined layering) rather than two-dimensional axisymmetric as modeled.

2) The predicted heater surface temperature histories match the measurements fairly closely (See Figure 3). Deviations noted include: a) early in time, the data lag the prediction by a few degrees; and b) after 180 days a marked increase in slope of the measured data occurs relative to the computed behavior.

3) Experimental data at and near the heater-hole wall, collected both parallel and perpendicular to strike, (Figure 4) display the same trend noted in the heater surface data. Rock surface temperature measurements lag the prediction early and diverge above the prediction late in time. Parallel to strike, temperatures at the 0.61 m radius show excellent correlation with calculations up to about 100 days. A gradual divergence of experimental data above the computed temperature begins after 100 days. Perpendicular to strike, data at the 0.6-m radius lie below the data parallel to strike. This may be due to anisotropic changes in conductivity caused by vaporization and/or mineral dewatering. Each of these phenomena occur over the temperature range 80-125°C. The hypothesis that changes induced by water loss near 100°C cause the temperature divergence between temperatures in S1-2 and S1-4 is strengthened by noting that the temperature data parallel and perpendicular to strike are nearly identical at all stations from 1.22-m radius outward (temperatures less than 85°C) and compare very closely with the computed temperature. It is noted, however, that at times greater than about 60 days the experimental data fall slightly below the computed results at the 1.22-m radius. This is opposite of the trends noted at the rock surface and 0.61-m radius.

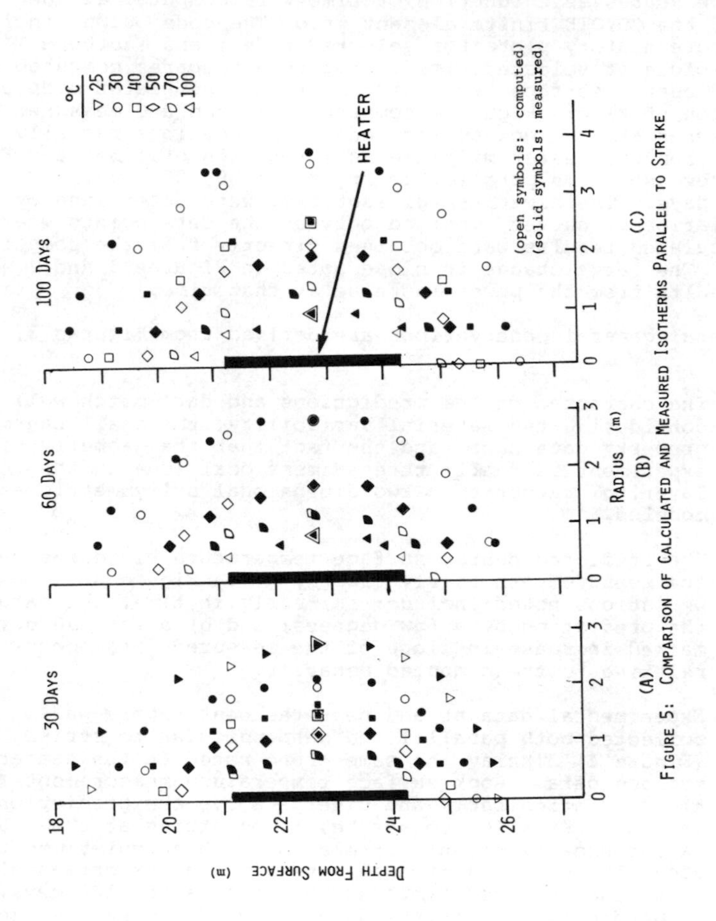

FIGURE 5: COMPARISON OF CALCULATED AND MEASURED ISOTHERMS PARALLEL TO STRIKE

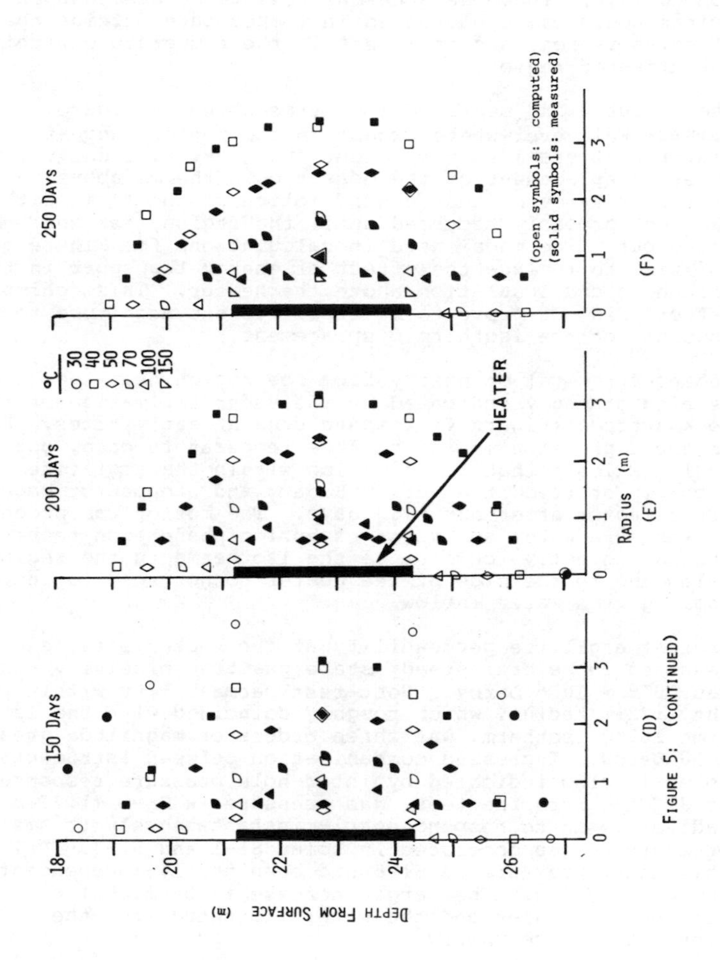

Figure 5: (continued)

The trend whereby the measured temperatures diverge above the calculations at 0.61-m location and below the calculation at the 1.22-m radius is consistent with a conductance decrease occurring in the volume within the 1.22-m radius. Note in Figure 5 (b and c) that the 1.22-m radius is at a temperature near 85°C, where phenomena than can influence the conductance, such as mineral dewatering, shrinkage and joint opening within the argillite, begin. When compared with temperatures for a region of non-altered conductivity, increased thermal resistance due to open joints would cause decreases in temperature outside the affected region, and increases in the temperature within the affected region.

4) The computed and measured isotherms shown in Figure 5 compare well everywhere except in the region above the heater at times less than about 100 days. The early time upward displacement of the measured isotherms above the heater is evidence that upward motion of the fluid within the rock probably occurred until the region near the heater dried out. The model used in calculations for Figure 5 was modified to include the effect of energy transport in the bore hole and insulation above the heater. This "chimney effect" did not prove to be of sufficient magnitude to account for the isotherm displacement.

   Enhanced removal of energy from the region near the heater is also strongly indicated by the under estimation of the rock surface (Figure 4) temperatures at early times. Based on the depression in the surface temperature data, one would estimate that fluid motion within the argillite probably started at about 5-20 days and stopped or became a minor factor after about 60 days. The motion was probably driven by a water and vapor expansion phenomenon rather than a convective cell since the isotherms in the region below and to the side of the heater do not seem to indicate cooling from water inflow.

   Pretest argillite permeability at the heater site, as measured in several steady-state gas transmissivity tests, was 10^{-3} - 10^{-2} Darcy. Post-test permeability within the 1.22-m radius, which roughly coincided with the late-time 100°C isotherm, was three orders of magnitude greater, 1-10 Darcy. Increased communication between istrumentation holes is also indicated by inter-hole pressure response. At 37 days into the test, gas pressure in S1-6 (1.22-m radius) began to respond nearly instantaneously to small pressure pulses generated in holes S1-1 and S1-3. Prior to this time, pressure in S1-6 had been nearly independent of the other holes. The large increase in permeability allowed pore water and the water liberated from the clay minerals to move upward.

Mechanical Model

A thermal stress analysis was also performed to predict the mechanical response of the Eleana argillite to the near-surface heater test. The predictions were used to help assess the argillite performance during the test and to aid in the evaluation of the predictive capability of the structural code using the laboratory property data.

The numerical calculations for the mechanical analysis were made with the ADINA [9] computer code. ADINA (Automatic Dynamic Incremental Nonlinear Analysis) is a general purpose, linear and nonlinear, finite element analysis program for static and dynamic problems. It contains an element library with 1-D, 2-D, and 3-D continuum finite elements, in addition to an extensive materials library. The present thermal stress problem is highly nonlinear, due to inelastic material behavior and temperature dependent properties; hence, the incremental solution for each thermal load step required several iterations.

Transient temperature distributions calculated with the COYOTE code, described in the previous section, were input to the stress calculations. Hence, the overall geometry and dimensions of the analysis region remained the same. The finite element configuration used for the structural analysis was different from that used for the thermal calculations. The mechanical model has 459 four-node quadrilateral elements with 500 nodal points. Since the finite element thermal and mechanical meshes were different, an interpolation program was used to transfer the temperatures from one network to the other.

Due to the proximity of the heater to the surface of the earth, initial in situ stresses were not considered in the mechanical analysis. The outer boundary (r = 13.5 m) of the axisymmetric geometry was assumed to be fixed with respect to radial (horizontal) displacements. The radius to this boundary was sufficiently large so that the boundary represented those conditions at infinity. Likewise, the boundary below the heater was assumed to be fixed with respect to axial (vertical) displacements. The top surface, and the borehole surface, were assumed to be traction-free.

Argillite in the Eleana Formation is a highly jointed rock mass, with average joint frequency of approximately 3-10 per metre near the surface, and 1-6 per metre at depths greater than 150 m [10]. Jointing at the near-surface heater site is representative of near-surface argillite. The jointing is multi-directional, and results in a rock mass composed entirely of small irregular blocks of intact rock, in immediate contact with each other. Core samples taken at various depths in the formation, especially near-surface, are observed to separate into several smaller pieces unless jacketed before laboratory testing. Because of the severely jointed and randomly oriented nature of the argillite rock mass, the material was assumed to be mechanically isotropic and homogeneous. It is apparent that the argillite rock mass possesses little tensile strength in situ. In compression, however, the argillite reveals a nearly constant elastic modulus at a given confining pressure, and a finite compressive strength, followed by strain softening. This overall material behavior is not unlike that exhibited by concrete, crushable foams, soils, and other geologic media.

In the ADINA code, the material constitutive description is input as a uniaxial stress-strain curve with strain softening. This is shown for argillite in Figure 6a. Data for argillite are discussed elsewhere [4]. No attempt has been made in these calculations to model the pressure or temperature dependence of the uniaxial stress-strain behavior.

The ADINA code employs failure surfaces to model tension or compression failure under multi-axial stress conditions. Based upon the current knowledge of Eleana argillite, the failure surface shown in Figure 6b has been used. Tensile failure occurs if the tensile stress in a principal stress direction exceeds the assumed tensile failure stress of 3.5 MPa. In this case, it is assumed that a plane of failure develops perpendicular to the principal stress direction. Once failure occurs, the normal and shear stiffnesses across the plane of failure are reduced, and the corresponding normal tension stress is released. Calculated

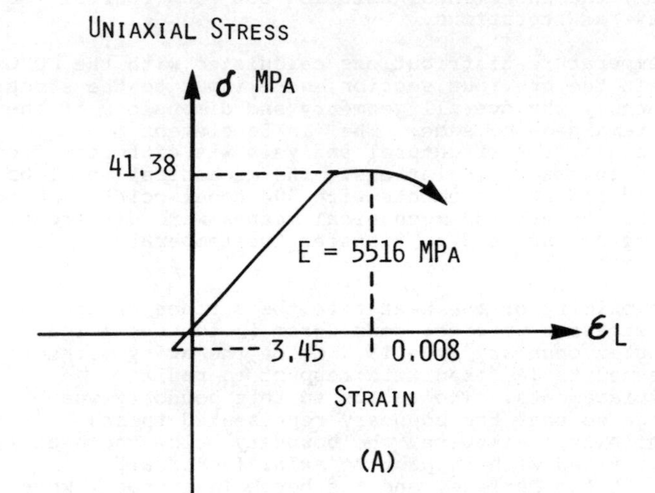

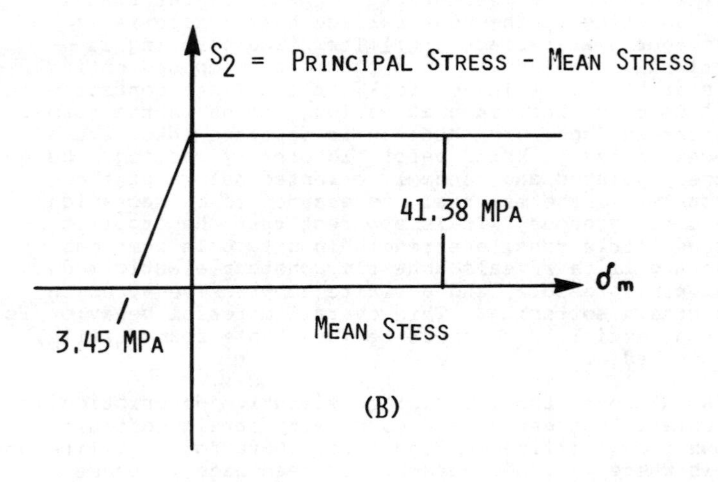

FIGURE 6: ASSUMED MECHANICAL PROPERTIES MODEL

(A) CONSTITUTIVE; (B) FAILURE

compression failure occurs if the deviatoric stress exceeds the compressive strength. This is the Tresca failure condition. Material property data defining the dependence of argillite compressive strength on confining pressure were lacking at the time of these calculations; hence, the compressive failure surface was assumed to be independent of the hydrostatic pressure. Although Cook and Witherspoon [11] recommend reducing mechanical strength properties of intact rock measured in the laboratory by a factor of 6 to 7 to obtain rock mass properties in situ, this was not done for Eleana argillite because the laboratory test specimens were already severely jointed, and compressive mechanical properties of argillite measured in the laboratory may be quite close to those of the in situ rock mass.

The combined effects of dehydration shrinkage and thermal dilatation for Eleana argillite used in this study are shown in Figure 7. Inherent in this data is the assumption that boiling occurs near 100°C, i.e., that the overburden stress does not effect the boiling temperatures and that the test is above the water table. Upon cool down, it is assumed that argillite, once dehydrated, does not rehydrate, and therefore does not follow in reverse the same expansion curve as on heating. Rather, with decreasing temperature, the argillite is assumed to follow its nominal linear thermal dilatation slope of 13.0×10^{-6}/°C. This behavior is also shown in Figure 7.

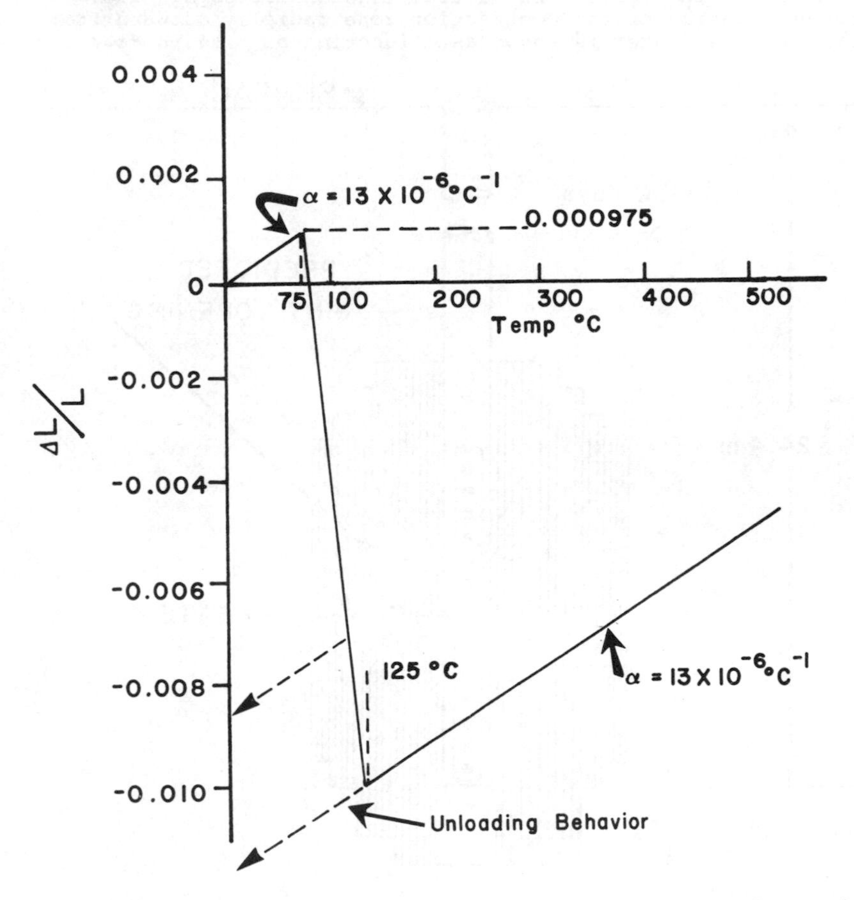

FIGURE 7: ASSUMED THERMAL EXPANSION BEHAVIOR OF ELEANA ARGILLITE

Mechanical Results

Upon non-uniform in-situ heating under the assumed conditions, the effect of the negative thermal expansion is a zone comprised of volumetric contraction, covering that portion of the argillite above approximately 75°C. Since the jointed argillaceous rock mass will not support tension, the deformation mechanism of interest in this zone is not compressive fracture of coherent blocks, but rather the contraction of intact jointed blocks and the opening of preexisting joints.

Regions in the r-z plane where the maximum principal stress has exceeded the tension failure stress of 3.5 MPa (see Figure 6b) are regions of joint opening in which joints are orientated tangent to cylindrical and conical planes. For the two-dimensional axisymmetric geometry under consideration, the circumferential stress is also a principal stress. If the circumferential stress has exceeded the same tension failure stress, then joints open along radial planes passing through the centerline of the axisymmetric geometry. In all cases, contours of the circumferential stress revealed the same regions of joint opening as did contours of the maximum principal stress in the r-z plane, indicating that the deformation mechanism was one of volumetric contraction. A contour plot defining the calculated zone of joint opening is shown in Figure 8 for a test time of 150 days. The computed propagation of the contraction zone radially outward from the center of the heater is shown as a function of time in Figure 9.

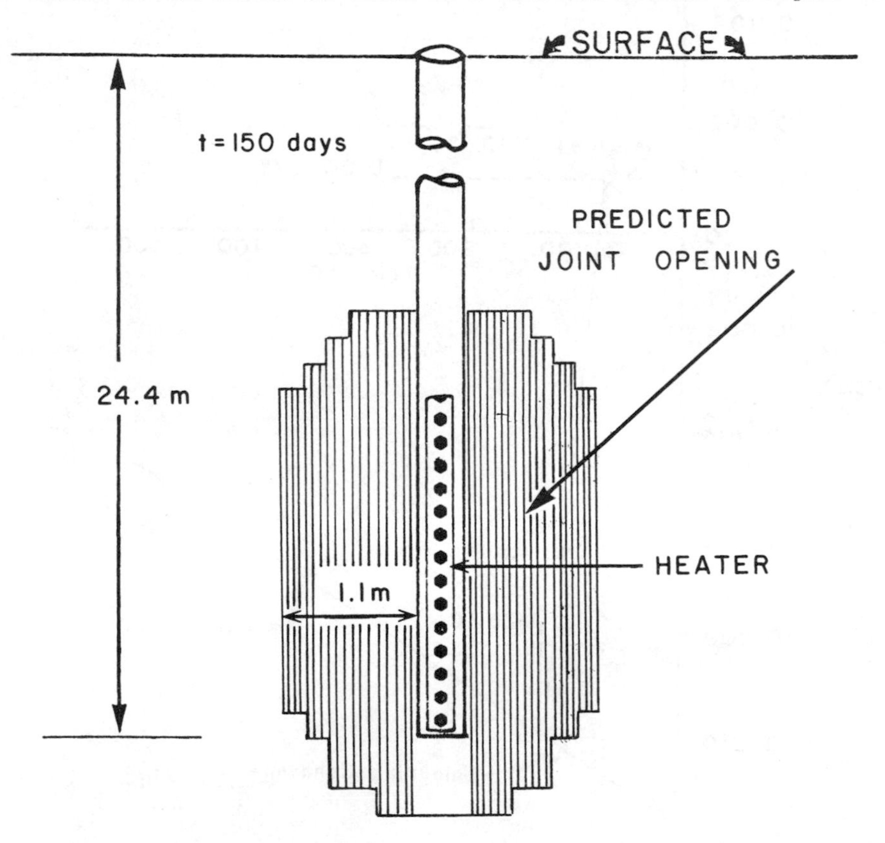

FIGURE 8: MECHANICAL MODELING RESULTS FOR FULL-SCALE HEATER TEST AT 150 DAYS

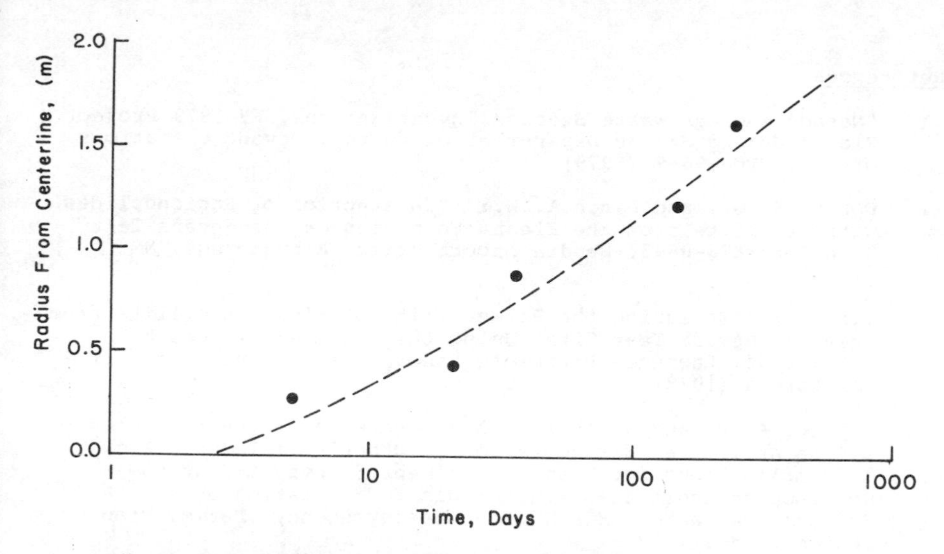

FIGURE 9: CALCULATED RADIAL EXTENT OF ZONE OF VOLUMETRIC CONTRACTION AT THE HEATER CENTER PLANE

It should be noted that, based on original assumptions, the calculated zones of volumetric contraction and joint opening do not disappear as the temperatures decrease to ambient from their peak values. In the Eleana full-scale heater test, the presence of joint opening zones after cooldown and removal of the heat source was confirmed by post-test heater hole inspection.

Conclusions

The Eleana full-scale near-surface heater test represents a first-cut, low-cost experimental effort at understanding the response of argillaceous rocks to the heating following emplacement of radioactive wastes. Though the near-surface environment is far from ideal for testing purposes, valuable insight was gained into material response and analysis techniques. Several conclusions may be drawn from the data and analysis:

First, it appears that the thermal and mechanical response of argillite is dominated by effects of clay contraction near the boiling point of water. This contraction causes the opening of preexisting joints in the rock, and results in:
1) greatly increased permeability resulting in increased transport of steam and water inside the 100°C isotherm; 2) the dominance of volumetric contraction rather than compressive forces near-field results in lack of compressive rock failure; and 3) the decrease of in situ thermal conductivity below laboratory-measured values.

Second, the thermal and mechanical models used for analysis of the heater experiment are in fairly good agreement with experimental results, though model development is obviously needed. Thermal modeling should be expanded to include treatment of anisotropies due to jointing and inclined layering (3-dimensional), as well as heat transfer within the high-permeability zone. Mechanical modeling should be expanded to treat preexisting joint systems.

Finally, although the thermal and mechanical interactions described here appear to lead to uncertainties in material response that can at present only be approximated, operation of the full-scale near-surface heater test in Eleana argillite identified no "failure" mechanism that might disqualify argillaceous rocks from further consideration as emplacement media for nuclear wastes.

References

1 "Nevada Nuclear Waste Storage Investigations, FY 1979 Project Plan," United States Department of Energy, Nevada Operations Office, NVO-196-9 (1979).

2 Dosch, R. G., and Lynch A. W.: "Interaction of Radionuclides with Argillite from the Eleana Formation on the Nevada Test Site," SAND78-0893, Sandia Laboratories, Albuquerque, NM (1979).

3 Lin, W.: "Measuring the Permeability of Eleana Argillite from Area 17, Nevada Test Site, Using the Transient Method," UCRL-52604, Lawrence Livermore Laboratories, Livermore, California (1978)

4 Lappin, A. R. and Olsson, W. A.: "Material Properties of Eleana Argillite, Extrapolation to Other Argillaceous Rocks, and Implications for Waste Management," Presented at the Workshop on Argillaceous Materials for Isolation of Radioactive Waste, OECD Nuclear Energy Agency, Paris, France, September 9-12, 1979.

5 Gartling, D. K.: "COYOTE - A Finite Element Computer Program for Non-Linear Heat Conduction Problems," SAND77-1132, Sandia Laboratories, Albuquerque, NM (1979).

6 Eckert, E. R. G., and Carlson, W. O.: "Natural Convection In An Air Layer Enclosed Between Two Vertical Plates with Different Temperatures." International Journal of Heat and Mass Transfer, 2, p. 106, (1961).

7 Touloukian, Y. S., and Dewitt, D. O.: Thermal Radiative Properties, Metallic Elements and Alloys, IFI/Plenum Press, New York (1972).

8 Lappin, A. R. and Cuderman, J. F.: "Eleana Argillite-Nevada Test Site," National Waste Thermal Storage Program Progress Report, 10/1/76 - 9/30/77, Office of Waste Isolation, Oak Ridge, Tennessee, Y/OWI-9, p. 223-230 (1978).

9 Bathe, K. J.: "ADINA, A Finite Element Program for Automatic Dynamic Incremental Nonlinear Analysis," Massachussetts Institute of Technology Report 82448-1 (1977).

10 Hodson, J. N., and Hoover, D. L.: "Geology of the UE17e Drill Hole, Area 17, Nevada Test Site," United States Geological Survey Open File Report USGS-1543-2(1979).

11 Cook, N. G. W., and Witherspoon, P. A.: "Mechanical and Thermal Design Considerations for Radioactive Waste Repositories in Hard Rock," LBL-7073, Lawrence Berkeley Laboratory, Berkeley, California (1978).

Discussion

J. MARTI, United Kingdom

As you have mentioned, there are strong interactions between the mechanical and thermal effects in an argillaceous repository. On the one hand, there are thermally-induced stresses which are not directly additive to other stresses if the material yields (or has otherwise non-linear behavior) ; on the hother hand, thermal phonomena are strongly affected by the stresses, for example through the changes in thermal conductivity accompanying cracking. Under those conditions, can an uncoupled analysis be justified ? Or, do you think that it is necessary to solve the coupled system of equations describing the stress-strain and thermal behaviour of the repository ?

A.R. LAPPIN, United States

You are absolutely right about the need for fully coupled analyses, and we currently have ongoing efforts aimed at being able to bring together thermal, mechanical and fluid transport analyses. For the Eleana experiment, however, we were limited by the amount of material property data we were able to collect, so that even available uncoupled codes were much more sophisticated than were the experimental material properties data we could put into them. Therefore, we used available codes, with little modifications. Ultimately, however, as you suggest, completely coupled calculations will be essential. It should be fun, given the large changes in permeability and thermal conductivity that appear to take place.

J. MARTI, United Kingdom

In view of the difficulties arising at high temperatures (chemical, thermal and mechanical aspects become far more complex), is there a realistic chance that those problems will be solved in the foreseable future or, rather,should we not concentrate in studying the material behaviour at lower temperatures (say, below 100°C) and provide some form of storage, prior to disposal, until the heat generation rate of the waste has decreased sufficiently.

A.R. LAPPIN, United States

I, personally, am concerned about the apparent uncertainties involved in going over 100°C. At this point, I do not feel we can model the response of argillaceous rocks above this temperature with much reliability at all ; nor do we know the consequences of the uncertainties with any degree of precision. So, yes, I agree that at this point in time, we should probably plan for long-term waste storage, long enough to ensure that temperatures resulting from emplacement would not exceed 100°C. In time, it may well be possible to understand the behavior above 100°C well enough to model it with sufficient accuracy, or to show that it has no bad effect on containment. At the present time we cannot do either one of these things.

D.F. McVEY, United States

There are at least two further reasons for investigating the material response at high temperature.

First, we will probably have to answer a series of "what if" questions when we apply to licence a repository. In order to give the public confidence in our statements we must be able to show we have done a thorough scientific investigation of all possible failure mechanisms. High temperature is one of these mechanisms.

Secondly, if we can demonstrate that we can design for a higher temperature it would probably result in considerable saving since the allowable power per unit area could be increased and the repository could be smaller.

N.A. CHAPMAN, United Kingdom

I have several questions linked to the local dehydration of the clay :

1) Do you have any estimate of the volume of water mobilised during the experiment ?

2) Where did it go ?

3) Would this fluid migration mechanism actually occur in a deep backfilled repository (would the expanding zone of contraction shown in Figure 9 be present and if so what would be its limits) ?

4) Finally, do you have any plans to exhume the near-field clay zone to examine morphological and mineralogical changes ?

A.R. LAPPIN, United States

1) Yes, we have made preliminary calculations of water evolution, based on the position of the 100°C isotherm, and I frankly do not remember the results now.

2) As I should have pointed out in the presentation, the heater hole extended below the bottom of the heater. Below the heater, this hole was cross-connected to a second hole, in which we placed a small sump pump. So, the water was simply removed after condensation for the first 50 days of the test. Though the system was fairly crude, it probably did remove any water that entered the zone below the heater until this time, in addition to any surface run-off. After about 50 days, we did not collect any more water, so we simply do not know where any water released by the rock after this time went.

3) Yes, I would expect it to happen at depth as well, so long as the rock was strong enough to sustain open joints. Certainly, it could happen during the operational phase. It could even take place after backfilling, unless the backfill had very low porosity and permeability.

4) Well, we do not have funding to exhume the entire zone. What we have done, though, is to drill two inclined drill holes from the surface back into the heater hole zone, one of which appears to have intersected the heater hole itself.

R. PUSCH, Sweden

All the difficulties associated with high temperatures led, very early, to a definition of a maximum allowable temperature of 100°C for the Swedish KBS concepts. The design later led to a maximum expected temperature of about 65°C. We agree with the previous speaker, that it is necessary, however, to test the concepts at much higher temperatures. This is supposed to take place in the Stripa mine in the framework of a cooperation project to which, among others, will participate a team of the Lawrence Berkeley Laboratory, California, led by Prof. Witherspoon.

A.R. LAPPIN, United States

One of the reasons we are concerned with the rock behavior at high temperature is that, because of public concern, we must always be able to answer the question "what if ?". So, even if we were to lower our design temperatures to 100°C, we simply must be able to explain what would happen in the event of some sort of "thermal runaway", where temperatures became much higher than we initially expected.

S. GONZALES, United States

Both the tests in the Eleana Argillite and the Conasauga Shale were conducted near the surface, and both rock units have experienced the influx of ground water. What would be your estimation of the water (as heated by a deep in situ test) effects, and related rock mechanical aspects, in a deeper argillaceous formation where ground water influx would not be such a factor ?

A.R. LAPPIN, United States

Actually, while there is no doubt about the presence of considerable ground water in the Conasauga Shale test, we have not been able to tell how much ground water entered the experimental site in the Eleana. Certainly, after about 50 days of heating, when we no longer saw any water collecting in the sump hole, there appeared to be no ground water effects. At depth, while there might not be any ground water problems, you would still have to account for and understand the behavior of the water released from the rock. So, I would expect the problem to decrease, but certainly not to disappear.

T.F. LOMENICK, United States

Were you able to find a test condition at the surface that is similar to the "target" horizon at depth ? Specifically were you able to find unweathered argillite with a high clay content ?

A.R. LAPPIN, United States

Yes and no. The horizon within which we carried out the test did have quartz and clay contents representative of Eleana argillite as a whole, and was below the depth of obvious alteration, i.e. the zone in which alteration and oxidation along joints was obvious. However, the test horizon does appear to be somewhat oxidized, since there are no sulfides present in it, in contrast to deeper materials.

GENERAL DISCUSSION

DISCUSSION GÉNÉRALE

D. RANCON, France

Au cours de cette séance, nous avons parlé de deux aspects, l'un géochimique, traitant de la rétention des radioéléments par les minéraux argileux, et l'autre, exposé dans les trois communications suivantes, sur les différentes modifications que peuvent subir les argiles sous l'influence de la température et de la pression.

Je pense qu'il faudrait peut-être relier ces deux parties. Ma question est donc quelle pourrait être l'influence de ces transformations sur la capacité de rétention des argiles. Y a-t-il ici des personnes qui ont fait des mesures dans ce sens ? Quelle est l'influence de la température sur la rétention des éléments, par exemple ? Comment la migration peut-elle se trouver changée en fonction des variations de température ou de pression in situ ? Ces sujets n'ont pas encore été abordés. En laboratoire, y aurait-il des études effectuées sur la rétention en fonction de la température, par exemple différentes argiles chauffées à des températures différentes, est-ce que ces propriétés sorbantes peuvent changer ?

R.H. HEREMANS, Belgique

Oui, Monsieur le Président, nous avons une série d'expériences en laboratoire sur des argiles qui ont été portées à des températures comprises entre 100 et 500°C. On ne peut pas dire que tous les résultats soient cohérents, mais d'une façon générale, dans les conditions de laboratoire, nous avons constaté une diminution de la valeur de Kd avec l'augmentation de la température. Cette variation des valeurs de Kd dépend du type d'éléments que nous avons étudiés. Il est différent pour le césium, le strontium et l'europium. Le maximum d'écart que nous avons constaté, c'est un ordre de grandeur environ, mais on ne peut pas dire que les études soient terminées.

D. RANCON, France

N'avez-vous pas observé dans certains cas des augmentations de Kd en fonction de la température ?

R.H. HEREMANS, Belgium

Non, peut-être exceptionnellement, mais sûrement pas systématiquement. Mais le maximum que nous ayons atteint au point de vue température, pour certains essais, est de 500°C. Nous n'avons pas été au-dessus.

D. RANCON, France

En Suède, n'y a-t-il pas eu des études effectuées sur la bentonite dans ce domaine ?

R. PUSCH, Sweden

I am not quite sure on what you want me to comment. Migration studies have been carried out in the laboratory in a series of tests where, I think, the temperature has been varied too. It turns out that these migration tests have to be performed with confined samples, in order to simulate the mechanical stress situation that exists in situ otherwise the density and therefore the permeability of the samples would be quite different.

G.D. BRUNTON, United States

Many years ago there was work done at the Oak Ridge National Laboratory on the effect of heat on clay. Dr. Tamura has reported on that and additional work has been done at Pennsylvania State University by Dr. Brinley. The experiments consisted of heating the clay to various temperatures and then measuring the ion exchange capacity after it had cooled down again. There was no determination at high temperature but they found that the exchange capacity of clay minerals varies with the temperature and with the element. For example cesium retention increases for certain minerals, sodium and calcium retention decreases. I am not aware of any more of that work being performed at the moment. In addition many years ago there was a proposal to incorporate waste radionuclides in clay minerals as a way of sealing them off. This study consisted of putting the elements on the clay minerals then heating the clay minerals up until they were no longer expanding or no longer had any exchange capacity. I think some of this work has been published in reports of the Oak Ridge National Laboratory.

D. RANCON, France

Je vois qu'il y a là encore un domaine à explorer, puisqu'il semble qu'il y ait assez peu de recherches qui aient été faites sur les mesures in situ dans les milieux chauffés ou sous pression. Je pensais que c'était plus avancé que ce n'est. Comme il nous reste un peu de temps, nous pouvons explorer d'autres sujets.

R. PUSCH, Sweden

I think that Mr. Tassoni's report on his field test and many other statements given today indicate the extreme difficulty of performing representative tests in natural argillaceous deposits. I wish to ask a question concerning the stratigraphical or structural occurrence of permeable units. Is it possible to identify and document the existence of permeable layers in sediments and, secondly, to identify the degree of continuity of such layers ? The same also goes for fissures and cracks. Being a soil engineer and working usually with soft clay I know that the thin permeable layers are of extreme importance since they control water movement and affect slope stability and associated phenomena. The soft clays which can be found in Sweden are considered rather homogeneous, but if you look at them in greater detail you find a lot of very thin layers, that cannot be detected in the field, but which must be of importance for the natural diffusion processes. Very often over-consolidated old clays are finely fissured, such as the Tertiary London and Oxford clays, so that experiments made in the laboratory give a certain permeability while in the field the fissure system is responsible for most water movement.

A.R. LAPPIN, United States

Detailed descriptions of argillaceous formations are difficult to find, probably because geologists have traditionally thought that clays were dull and have not really looked at them. Therefore the data base on the mineralogical and mechanical properties of argillaceous formations is quite limited. In relation to the correlation and lateral extent of joints we are doing some work in welded tuffs. We hope to set up experiments where we can look at controlled isotope migration in the field along a well-defined joint. As the first step parallel holes about 1 metre apart have been drilled ; the only joints which could be correlated with any regularity were either tectonic or appeared to be due to elastic unloading very near to the tunnel wall. The cooling joints and other joints in the tuffs could not be correlated over distances of three feet. If you are going to run a radionuclide migration experiment in situ you should be able to tell the health physics people that you know where the tracers are likely to go. In tuff, at this point, it would be difficult to make reliable predictions.

C.R. WILSON, United States

I agree that characterizing the hydrology of argillaceous sediments may be a difficult problem. On the basis of my experience with near surface clays, identifying areas where there are no sand lenses or joints could be very difficult in certain formations. These things seem to occur almost randomly and it is very difficult to predict them. You may be in a perfectly solid thick layer of clay in one place but a few meters away you may have a large isolated sand lense. Extensive exploratory drilling could provide the detailed information but it would also destroy the integrity of the formation and reduce its effectiveness as a containment barrier. It may be necessary to limit our explorations to those clay layers which do not contain many sand lenses and proceed on a statistical basis.

A.A. BONNE, Belgium

I should like to give some information about work done in Mol on radionuclides migration. A model has been developped ; this takes into account diffusivity and permeability in clay ; the result of the sensitivity study has been that the permeability of the clay is not the factor controlling migration. Diffusivity is the most important factor.

J. ROCHON, France

Justement, ne pensez-vous pas qu'une variation de diffusibilité avec la température peut introduire une variation de perméabilité ? Je ne le sais pas dans le cas des argiles, mais nous avons fait des expériences sur des calcaires. Ces expériences montrent que, dans le cadre de l'utilisation de la chaleur géothermique, lorsqu'on réinjecte de l'eau froide, le calcaire se dissout, et ensuite l'eau se réchauffant, il y a précipitation du calcaire dissout dans l'aquifère, donc variation de perméabilité. En fait, tous les facteurs chimiques, thermiques et hydrauliques s'entrecroisent, et je crois que c'est là le problème à résoudre.

D. RANCON, France

Plus de commentaires ni de questions ? Je crois que nous allons conclure cette séance. On peut remarquer pour conclure, qu'il y a encore un immense problème à étudier, et que lorsqu'on va passer aux études in situ de migration, qu'il y aura encore beaucoup de travail à faire.

Session 3

Chairman-Président
M. R.H. HEREMANS
(Belgium)

Séance 3

AN INVESTIGATION OF THE REGIONAL CHARACTERISTICS OF THE DEVONIAN SHALES FOR THE STORAGE/DISPOSAL OF RADIOACTIVE WASTES

by

T. F. Lomenick
Oak Ridge National Laboratory
Oak Ridge, Tennessee 37830

R. B. Laughon
Battelle Memorial Institute
Office of Nuclear Waste Isolation
Columbus, Ohio 43201

ABSTRACT

Devonian-age shales underlie a large portion of the northeastern quadrant of the country and are abundantly thick and relatively near to the land surface within the Appalachian, Illinois and Michigan basins of that area. Although these thick accumulations of argillaceous sediments are best-known for their natural gas production and potential, they would also appear to be good candidate rocks for radioactive waste repositories as the shale is generally impermeable, and therefore, relatively free of circulating ground waters to disperse any waste emplacements.

Introduction

Although the research, development and demonstration (RD&D) program in the United States for the permanent storage of high-level radioactive wastes has focused its efforts on bedded-salt formations, the potential suitability of other subsurface rock types, such as granite, basalt and argillaceous sediment, has long been recognized. With detailed investigations, these alternatives may, in fact, be shown to offer comparable physical, chemical and geological properties.

Under the Office of Nuclear Waste Isolation (ONWI), which is being administered by the Battelle Columbus Laboratory for the U.S. Department of Energy (DOE), work is being directed toward efforts to more fully investigate some of the most promising alternatives to rock salt. One of these alternatives, which is the subject of this paper, is being examined by the Oak Ridge National Laboratory as an integral feature of ONWI's Terminal Waste Storage Program. Although argillaceous rocks are widespread throughout the continental U.S., the Devonian shales that underlie a large portion of the northeast are particularly attractive for study.

Firstly, because the Devonian black shales are petroliferous, they are presently being extensively investigated for their natural gas production and kerogen conversion potential.[1] Since the early 1800's, high quality gas has been produced in modest amounts for local domestic and small commercial markets; oil production in these same general areas, however, has been primarily realized from siltstones and sandstones that overlay or feather into the black shales. The determination of the Devonian shale hydrocarbon potential is presently being carried out through DOE's Eastern Gas Shales Project, which is being administered by the Morgantown Energy Technology Center, Morgantown, West Virginia.[1] This work is providing extensive characterization data of these Devonian shales, which include statigraphy, mineralogy, porosity, permeability, structure, etc. Even though some shale characteristics that would be favorable for gas production might not be acceptable features for a radioactive waste repository, e.g., a high degree of fracturing, it is clear that most of these data would be directly applicable to the assessment of both potential uses. Consequently, near-term research results can be achieved for argillaceous rocks for the Terminal Waste Storage Program by using this available data base and technology that are being collected and developed specifically for these Devonian shales.

Secondly, the Devonian shales were selected for investigation by OWNI when earlier scoping analyses pointed to favorable conditions for more detailed study. This summarized the geologic and hydrologic characteristics of selected salt and impermeable sedimentary rock deposits, including argillaceous formations, for the purpose of identifying potentially suitable repositories for deep geologic disposal.[2] Many shale deposits in the U.S. could not meet the selection criteria established for the study. The primary reasons for this tentative rejection of shale formations included high seismic risk; relatively high permeability; inadequate thickness or excessive depth; interbedding with permeable strata of siltstones, sandstones and limestones; and extensive fracture and deformation due to regional tectonism. However, many shale deposits in the U.S. appeared to have characteristics that made them worthy of more detailed investigation. Three of these shale formations are located in the northeastern U.S. and include (a) the Devonian-Mississippian Ellsworth Shale in Michigan; (b) the Mississippian Coldwater Shale in Michigan; and (c) the Devonian Ohio Shale in central and northern Ohio. These shales lie in the basins being examined by the Eastern Gas Shales Project.

Characterization of Devonian Black Shales

Those argillaceous sediments of Upper Devonian and Mississippian age that are commonly referred to as black shales are found in twenty-six states in the U.S. and in extensive portions of Canada.

The terminology "Devonian black shale" is generally used in references to a sequence of contiguous fissile shales, mudstones and claystones, including any interbedded foreign strata such as siltstones and sandstones. Consequently, as would be expected, black shale formations can exhibit more or less a single color, or can show marked variations in coloration, and in either case, dark or black beds may or may not be present. The colorations found may vary from light shades of grays, greens and browns to dark and blackish values. In addition to similar age and clay mineral composition, an important feature common to all of the so-called black shales is their high organic content. While variations in coloration can be partially attributable to differences in the sources of the clastics and their nonclay constitutents, the colors shown by these argillaceous sediments are likely to also reflect the thermal maturity of the carbonaceous content. The black shales under investigation by the Eastern Gas Shales Project are located in the Appalachian, Illinois and Michigan basins. The areal extent of these three basins, and their corresponding massive argillaceous sediments, is a significant portion of the northeast quadrant of the U.S.

In the Appalachian Basin, which reaches from southwestern New York to northern Alabama, the Ohio and Chattanooga shales are the two best known shale formations. The Ohio Shale, which is often referred to as the Ohio Brown Shale, has been the dominant source of natural gas production in these three basins since the early 1800's, and consequently, the Ohio Shale and northern portions of the Chattanooga Shale are the eastern shales of most interest for stimulated natural gas production. The western boundary of the Appalachian Basin corresponds to the western outcrop of the Ohio Shale. This north-south outcrop extends for the full length of the state and into Kentucky. There are three ascending Devonian shale members in the Ohio Shale sequence, which are known as the Huron, Chagrin and Cleveland Shales. The Devonian-Mississippian Bedford Shale and Lower Mississippian Berea Sandstone formations rest on the Ohio Shale. The Ohio Shale sequence is brownish to grayish-black in appearance and varies in thickness from about 400 ft in central Ohio to more than 4000 ft in the southeastern part of the state.

The Devonian black shales in the Illinois Basin which extends through major portions of Illinois, Indiana, Kentucky and Tennessee, are represented by the New Albany Shale sequence. In ascending order, the New Albany Shale consists of the Blocher, Sweetland Creek, Grassy Creek and Hannibal Shale members. The coloration in the sequence varies from greenish-gray to very dark browns and black. The brownish-black carbonaceous shale constitutes more than 80% of the New Albany in the southwestern part of the basin, but elsewhere its abundance declines to 20 to 40 percent. The thickness of the New Albany Shale sequence varies from about 100 ft near its outcrop at the Cincinnati Arch to more than 300 ft in extreme southwestern Indiana. The New Albany Shale both thickens and dips into the Illinois Basin. Compared to the Appalachian Basin, natural gas production from Indiana's New Albany Shale has been modest. Although gas has been produced from about ten small fields in the past, all but one have been abandoned. From these New Albany gas fields, little correlation has been observed between thickness of the black shale and the occurrence of gas, even though these are organically rich shales.

The Devonian black shales in the Michigan Basin consist of a three member shale sequence which is commonly referred to as the Antrim Shale. Although the Michigan Basin is almost entirely limited in areal extent to the southern peninsula of Michigan State, the basin and Antrim Shale extend a short distance into northern Indiana and the northwestern extreme of Ohio. In ascending order, the three members of the Antrim Shale are the Antrim, Ellsworth and Sunbury Shales. Antrim shale exhibits beds of both black organic rich deposits and greenish-gray calcareous shale and the Ellsworth consists of interbedded greenish-gray nonorganic shale and black carbonaceous shale. The Sunbury is a black organic shale that shows a high gamma trace, but this stratum is quite thin. The Antrim Shale sequence lies beneath the Mississippian Coldwater Shale formation, and in places where the thin black Sunbury Shale is absent, the Ellsworth cannot be readily distinguished from the overlying Coldwater. Although there have been reports of gas showings in the Antrim Shale sequence, there are no commercial wells in the Michigan Basin.

Terminal Radioactive Waste Storage Program

With the above introductory information and brief characterization of the Devonian black shales as background, the following is a summary statement of OWNI's ongoing and planned research program for the deep geologic disposal of radioactive wastes in these rocks. This research will focus on the argillaceous rock formations in the Appalachian, Illinois and Michigan basins, which are also considered to be promising for their hydrocarbon potential. Because of the natural gas stimulation and kerogen conversion potential of the shales in these basins, the DOE and the oil and gas industry have been carrying out an active RD&D program to better understand the nature of the argillaceous deposits. While it is clear that some of the preferred rock characteristics for gas and in situ shale oil production, e.g., high permeability and organic content, are contrary to those rock qualities sought for a radioactive waste repository, it is recognized that a cooperative research effort between these two DOE programs can have a mutually beneficial and synergistic effect.

The ONWI-ORNL RD&D program pertaining to argillaceous sediments is in its initial research stage. The thrust of this research activity is to determine the suitability of using massive fissile shale, mudstone and claystone formations, which can be anticipated to show variable interbedding with siltstone, sandstone, limestone and dolomite, as a geologic setting for a terminal radioactive waste repository. At first, the results of this work will include lithological, mineralogical, geochemical, geophysical, hydrological and engineering investigations based on the published literature, the open files of the Federal and state Geological Surveys and the data base and technological advancements provided by the Eastern Gas Shales Project. During this first phase effort, no field studies or tests on the black shales are expected to be performed by ONWI-ORNL. This is because there already exists a large foundation of data and technology that is readily available from extensive field measurements made by others. For example, during the October 1978 Second Eastern Gas Shales Symposium, the number of core locations reported for the Eastern Gas Shales Project alone were 19 wells cored, 13 planned, and 29 proposed.[1]

Hence, this initial shale study will involve the analysis and interpretation of all the relevant data and information that can be assembled. These data have resulted from shale characterization work, which includes lithological, chemical, physical and mineralogical parameters, and well logging, core analyses and fracture and seismic determinations based on various geophysical and remote sensing techniques. It is intended that the results of this first phase investigation by ONWI-ORNL on shales will considerably advance our understanding of the Upper Devonian-Mississippian black shale sequences in the Appalachian, Illinois and Michigan basins in the following areas:

a. the statigraphy of individual strata and their interbasin correlations, including imperfections in the massive argillaceous sediments due to interbedded permeable sandstones, limestones, etc.

b. structure,

c. mineralogy, including the relationships between color, gamma logs, concentrations of uranium, thorium and other metals such as nickel and vanadium, and organic carbon content,

d. resource potential, with emphasis on locations exhibiting a low potential for natural gas stimulation, kerogen conversion and coal production,

e. the porosity, degree of impermeability and natural fracturing and hydraulic fracturing characteristics,

f. seismic risk and other tectonic considerations, e.g., folding, deformation, and

g. groundwater hydrology and the accounting of all boreholes.

It is anticipated that the results of this first phase effort will provide the means to develop criteria to be used in subsequent and more detailed investigations. Such criteria would be necessary for the identification of specific formations and geographical areas that merit further evaluation, which might involve well coring and logging, core analyses and other field determinations.

References

1. Overbey, William F., Jr., "Present Status of the Eastern Gas Shales Project," Second Eastern Gas Shales Symposium, METC/SP-78/6, Vol. II, U.S. DOE, Morgantown Energy Technology Center, Morgantown, West Virginia, October 1978.

2. "Shales, Mudstone and Claystone as Potential Host Rocks for Underground Emplacement of Waste," U.S. Geological Survey, Denver, U.S.G.S.-4339-5, Open-file Report, 1973.

Discussion

R.H. HEREMANS, Belgium

Comparé au sel et à la roche dure, quel est l'effort financier consenti pour la recherche sur matériaux argileux aux Etats-Unis ?

T.F. LOMENICK, United States

Perhaps the funding for shale related studies in the United States for the current year would total less than 1 million dollars or only a few percent of the total geologic disposal budget. Higher funding levels are planned in the future.

R.H. HEREMANS, Belgium

L'objectif du programme sur les schistes aux Etats-Unis est-il d'identifier un ou plusieurs sites potentiels de rejet ?

T.F. LOMENICK, United States

No specific date has been set for selecting a site for a repository in shale. However, after the first repository site in salt rocks has been identified, it is likely that a site in shale or granite will follow.

R.H. HEREMANS, Belgium

Dans l'affirmative, quand envisage-t-on pouvoir débuter des travaux sur le site sélectionné ?

T.F. LOMENICK, United States

Specific plans for further in situ tests in shales are not finalized.

F. GERA, NEA

Are direct observations of permeability and ground water flow available in quarries, mines or wells in some parts of the Devonian shales ?

T.F. LOMENICK, United States

At this time we are not aware of such direct measurements of permeability in mines, wells etc. However, measurements of permeability have been made on selected samples of core and they generally fall in the micro darcy to milli darcy range.

A.R. LAPPIN, United States

I am familiar with the fact that one concern in the WIPP project in southeastern New Mexico is possible drilling through the repository in search of hydrocarbons that might exist at greater depths. Is there any significant production of hydrocarbons from beds underlying the shales you are examining in the three basins ?

T.F. LOMENICK, United States

In general, most of the oil and gas production in the northeast is from rocks that lie above the black shale.

S. GONZALES, United States

I wish to comment about Dr. Gera's question about natural permeability and hydrology of Devonian Shales : some hydrocarbon (gas) production has been realized from these shales through natural fractures, especially in some older fields in the Appalachian Basin. As a matter of fact one focus in the Eastern Gas Shales Project is artificial (hydraulic fracturing) stimulation. In certain areas, there has also been success from remote sensing data to utilize fracture pattern analysis to locale more productive (i.e. gas) zones due to naturally permeable fractures.

I have another comment in relation to Mr. Lappin's question about hydrocarbon production from zones deeper than the Devonian shales in the eastern basins : localized hydrocarbon production from the Silurian Clinton sands has been established, namely within the Appalachian Basin, especially in eastern Ohio. In all basins, deeper exploration boreholes have been drilled below the Devonian, but most have been unproductive. Deeper production from the Silurian and Ordovician sequences has been established in part of the Michigan Basin.

THE CONASAUGA NEAR SURFACE HEATER EXPERIMENT-- IMPLICATIONS FOR A REPOSITORY SITED IN A WATER-SATURATED ARGILLACEOUS FORMATION*

J. L. Krumhansl and W. D. Sundberg
Sandia Laboratories**
Albuquerque, New Mexico
U.S.A. 87185

ABSTRACT

Results of the Conasauga Near Surface Heater Experiment suggest that the effects of ground water on a repository located in shale would be most pronounced in their influence on chemical processes. Despite the water-saturated condition of the Conasauga shale, conduction, not convection, was the principal mode of heat transport. Metal corrosion, however, was found to be significantly enhanced by the refluxing steam. The several centimeters of scale left behind by the boiling water, in a repository, could interfere with retrieval operations or form a coating of low thermal conductivity material on a canister. Finally, alternate wetting and drying due to sporadic steam generation caused up to 20 microns of alteration on a sample of borosilicate waste glass simulant (PNL 75-25) included in the test.

*This work was supported by the U.S. Department of Energy under contract DE-AC04-76DP00789.

**A U.S. Department of Energy facility.

Three technical concerns that are common to siting of a high-level waste repository in a geologic medium are presented in this paper. They are the effects the waste will have on the repository media, whether the packaging material(s) will retain their integrity during or following emplacement, and finally, whether the waste form itself will be stable over the long term in the repository environment. Each of these issues is a complex problem deserving of substantial laboratory and modeling programs. At some point, however, it also becomes necessary to abandon the conceptual pictures that have evolved regarding the probable behavior within a repository and deal simultaneously with all the complexities inherent in natural settings. Near-surface heater tests are a first step in this direction. Such tests assess the effects of heat alone without radiation, and are carried out at a shallow depth rather than one appropriate for an actual repository. Data from such experiments provide a preliminary assessment of whether laboratory and modeling programs are focused on the processes which will, in fact, dominate the behavior of the system, and also provide an opportunity to observe whether existing models are capable of dealing with geologic materials even in the more easily interpreted near surface environment.

The Conasauga Near Surface Heater Test is one of two experiments the United States has operated in argillaceous rocks. The formation chosen for this experiment consists predominantly of a friable silty illitic shale. Bedding is pronounced and the shale is interspersed with numerous layers of limestone and siltstone. The formation has also been sheared in a variety of directions as a consequence of the tectonic activity responsible for the Appalachian Mountains. Rainfall in the region averages 1.7 m per year; consequently, water is constantly available to refill the network of hairline fractures that pervade the formation in the subsurface.

The setting of the experiment is illustrated schematically in Fig. 1. Around each of the 3 m long by 0.3 m diameter heaters, S1-1 and S2-1, is an array of thermocouples, horizontal stress gauges and vertical extensometers emplaced at various depths in drill holes. Both heaters were operated with a midplane temperature of 385°C. At Site Two, an external air pressure of 0.19 MPa was applied, which prevented ground water from flowing into the site. At Site One, only 0.08 MPa was applied, barely preventing inundation of the heater.

The thermal loadings presently being proposed for canisters of high-level waste at reasonable canister spacings will not cause clay minerals such as occur in the Conasauga, Table I, to invert to more stable phases. Consequently, much of the concern regarding the permissible thermal loading of a repository in argillaceous rocks centers on the effects of dehydration, ground water circulation, and to a lesser degree, on the consequence of oxidation of ferrous iron compounds. The potential consequences of dehydrating samples of Conasauga Shale are illustrated by two simple laboratory experiments. A marked thermal contraction, Fig. 2, accompanies drying of unconfined samples in the laboratory. This contraction may cause a myriad of tiny cracks to open, resulting in a marked drop in thermal conductivity, and interfering with heat transport, Fig. 3.

The actual response of the formation during the test was somewhat different than might be anticipated from the response observed in the laboratory. No evidence was found for fracturing of the formation either in regions where the temperatures were above or below the boiling temperature of water. After the test, the annular marks made during drilling were still visible in the hole walls and the absence of large accumulations of shale fragments on top or along side of the heaters confirmed that small scale fracturing and spalling of the walls had not occurred to a significant degree. By the same token, it was also apparent that no large fractures opened in the hole walls during heating. Gas permeability measurements made between injection and exit wells several feet apart showed that, if anything, the bulk of the formation experienced a permeability decrease. Core samples recovered after the test confirm that prevalent fracturing was not caused by the heating cycle. The only textural changes that were discernible were a discoloration due to iron oxidation and a subtle loss of sheen on some slickensided shear surfaces. It is inferred that even a modest amount of confinement prevented the opening of a network of hairline fractures, and that the pervasively sheared nature of the formation relieved stresses locally to the extent that new large fractures were not formed.

TABLE I

Mineralogy of Conasauga Samples

	Calcite	Feldspar	Quartz	Kaolinite	Illite	Chlorite
Limestone	Major	Minor	Minor	-	-	-
Vein Calcite	Major	Minor	Minor	-	-	-
Gray-Green Siltstone	Minor	Major	Major	Minor	Trace	Minor
Gray-Green Shale	Minor	Major	Major	Minor	Major	Minor
Maroon Shale	-	Minor	Minor	Minor	Major	Trace
Dark Gray Shale	Minor	Minor	Minor	Minor	Major	Minor

The lack of in situ fracturing is reflected in the formation's thermal response. In contrast to the laboratory thermal conductivity results, the formation retained an overall conductivity in the vicinity of 1.8 W/m°C during the heating cycle. The above value was arrived at by assuming the conductivity to be temperature independent and that convective transport of heat was insignificant. Heater wattage measured in the field was then matched with values predicted by a numerical model, CINDA [1]. The next step in the analysis was to determine how a temperature dependent thermal conductivity would alter the above conclusions. As indicated by Fig. 3, the decreased conductivity measured in the laboratory occurred with the loss of water; consequently, the somewhat arbitrary selection of 120°C was made for the temperature at which the modeled conductivity would go from the higher value characteristic of lower temperatures to that of the lower value postulated to exist once dehydration had occurred. Two pairs of conductivity values, 1.5 and 2.0, and 1.75 and 2.25 W/m°C, were analyzed. In both cases, it was found that the modeled heater wattages fell close to the value that would have been calculated assuming a bulk conductivity for the formation equal to the average of the two chosen conductivities. Because of the possibility that the average conductivity may reflect a marked decrease at lower temperatures coupled with an increase at higher temperatures, it is necessary to get an independent estimate of either the high or low temperature thermal conductivity. Treating the heater as a line source, it follows from measured values of the heat production per unit length, Q/ℓ, the temperature gradient, $\partial T/\partial r$, at a radius r, and the formula for steady state heat conduction,

$$\frac{Q/\ell \ \ln \frac{r_1}{r_2}}{2\pi(T_1 - T_2)} = \kappa \quad ,$$

that a value of 1.8 W/m°C is indicated for the high temperature conductivity, κ. It is apparent that a precipitous drop in conductivity did not occur with the in situ dehydration of the Conasauga shale, and, therefore, that an analysis based on a constant conductivity is a reasonable model of the actual field situation.

So far, the analysis has assumed that the presence of ground water had a negligible effect on heat dissipation within the formation. It is possible to assess the validity of this assumption, but first it must be shown that the field isotherms, in fact, represent the total energy being put into the formation by the heater. That is, a case must be made for the fact that ground water flowing through the site is not carrying away a significant fraction of the heater input. For the first fifteen weeks of the experiment, the entire heated region was within the thermocouple array; consequently, it is possible to do an energy balance, by comparing the integrated heater output with the energy required to account for the observed temperature profile (see Table II).

TABLE II

Energy Input at Site One (x 10^{10} Joules)

	Computed From Heater Power	Computed From Isotherms
30 days	1.8	2.0
60 days	3.4	3.2
150 days	5.6	5.3

Since agreement is good, it follows that inferences based on isothermal positions and heater wattages are, in fact, a valid reflection of the formation's total response.

Given that ground water as occurs in the Conasauga will not serve to dissipate widely the heat from a waste canister, it is still of concern whether this water will greatly perturb the isotherms in the immediate vicinity of a single waste canister. Fig. 4 compares the actual and computed isotherms after 135 days assuming a temperature independent conductivity of 2.0 W/m^{o}C. Since conductive heat flow occurs perpendicular to isotherms, it follows that in the region of the heater midplane, the heat flux is radial as previously assumed, and that convecting water or steam is not having a significant effect on local heat dissipation in this region. In the far field at temperatures below 100°C, there is also good spatial agreement between computed and actual isotherm positions, indicating again that conduction is the predominant mode of heat transfer. There is, however, a perceptible displacement of the isotherms in the near field below and above the canister, indicating that ground water below the canister constituted a heat sink and that there was upward transport of heat due to steam circulating above the heaters.

Modeling of an open heater hole in a porous medium is, unfortunately, beyond the present state of the art. It is, however, possible to model the effects of a heater buried in a porous medium without the presence of a heater hole. The patterns of both water movement and heat flux predicted by this model, assuming a permeability of 100 milli-darcys, are qualitatively the same as observed in the Conasauga test. Since the geometry modeled is a reasonable approximation to a waste canister grouted in place, it follows that in a repository located in a medium similar to the Conasauga's, ground water circulation would influence heat dissipation in a manner similar to that documented for this field test.

The influence of water may extend beyond its effect on the heat transfer properties of the formation. The ability of many canister materials to withstand high temperatures is markedly influenced by the presence of water. As mentioned above, the two Conasauga heaters were operated under slightly different conditions. At Site One, the bottom of the heater was immersed in ground water for much of the test. At Site Two, sufficient external air pressure was applied to the heater hole to prevent ground water from flowing into the heater hole. Consequently, the only moisture in the heater hole originated as pore water in the shale. At Site Two, a uniform reddish-brown discoloration of the 304-stainless steel heater was the only corrosion visible on the heater. A suite of metallurgy coupons affixed to the heater base also showed negligible corrosion. At Site One, corrosion was more pronounced. Small blotches of intensified corrosion were noted on the top of the heater where abundant steam apparently condensed and collected. More spectacular, however, was the stress corrosion cracking of a string of thermocouples sheathed in 304-stainless steel. For a distance of roughly 0.6 metres above the heater, the sheath was totally destroyed. Thus, it was not the highest temperature regions which experienced the greatest corrosion, but the region where steam condensed and liquid water collected on metal fittings. It follows that the effects of refluxing steam may seriously affect canister life.

The lower part of the Site One heater was protected by scale deposited from the boiling ground water. The scale consisted of a porous mixture of predominantly anhydrite with lesser amounts of calcite, gypsum, and silica. Beneath the thicker portions of the scale, the heater was generally unrusted, and the only evidence of corrosion was an occasional gray-green discoloration on the inner surface of the scale. In spite of its beneficial effect on the corrosion rates, such scale could constitute a serious hazard in an actual repository. Up to 2.5 cm of scale was deposited on a three inch collar beneath the heater. By virtue of its porous nature, such an accumulation on the side of a canister could act as an insulator causing waste temperatures to reach values far higher than anticipated based on formation thermal conductivities. A second matter is that of retrievability. In trying to recover the Site One heater, it was found that the boot of scale had securely cemented the heater to the hole bottom.

If all other confinement systems fail, the integrity of a repository may depend on the stability of the waste form itself. In a water-free environment, borosilicate glass waste forms may be expected to devitrify rapidly only at temperatures considerably in excess of those presently being planned in a high-level waste repository. There is, however, a considerable body of literature indicating that both natural and artificial glasses alter considerably more rapidly in the presence of moisture. As a consequence, a sample of PNL 75-25 waste glass simulant was sandwiched between two pieces of Conasauga core and suspended at a depth of roughly 10.0 metres in a drill hole 1.3 metres from the Site One heater hole. During much of the test, the temperatures oscillated around 100°C due to the refluxing of steam from the lower parts of the hole.

Alteration of the glass over the 243 day extent of the test was perceptible. Thin layers of a yellowish gel formed having an aggregate thickness of about 20 microns. Elements such as Zn, Ba, and Gd were found throughout the gel, demonstrating that the gel was, in fact, an alteration product of the glass rather than a precipitate from the condensate that periodically bathed the sample. Although interaction with the rock was not perceptible, the gel itself had crystallized on its exterior surface to form a mat of acicular crystals. X-ray diffraction of the gel plus crystals gave lines in order of decreasing intensity at 4.3, 7.6, 3.3, 2.7, and 2.9 Å; however, no phases could be positively identified. Beneath the gel, the glass appeared to be fresh, but the surface was scalloped in a manner reflecting the shrinkage cracks that had formed during various stages of alteration. Thus, while formation of a surface coating may slow glass alteration, it is evident that alternate drying and wetting can mitigate the effectiveness of this barrier.

To summarize, in spite of its near surface environment, a variety of observations were made during the test that have bearing either on repository construction or on fielding of a pre-repository vault type test. Most chemical processes will operate in a similar manner either at depth or in a near surface environment. Furthermore, in argillaceous rocks, it will never be possible to preclude completely the presence of ground water. It follows that the corrosive nature of refluxing steam, the consequences of scale deposition, and the effects of alternate wetting and drying that were noted during this experiment all are pertinent to the design of a repository. It is also highly likely that in a volume of rock large enough to encompass a repository that some regions will already be fractured or develop fractures as a consequence of repository construction or use. An estimate of the perturbation that such a change would introduce may be gained, however, by considering that the data reported here reflects the thermal and mechanical responses of a somewhat fractured, water-saturated rock. In general, thermal conductivities remained high and were not decreased significantly by dehydration effects. This was, in turn, a reflection of the fact that rather than fracturing further as a consequence of being heated, the formation decreased in permeability. In short, the thermal and mechanical data obtained from this experiment reflect favorably on shale as a medium for high-level waste disposal. However, a variety of localized, non-equilibrium processes were noted that could operate in essentially every geologic setting proposed for the disposal of nuclear waste, and which could severely compromise the integrity of a series of engineered barriers.

REFERENCES

1. Chrysler Improved Numerical Differencing Analyzer TN-AP-67-287, Chrysler Corporation Space Division, October 1967.

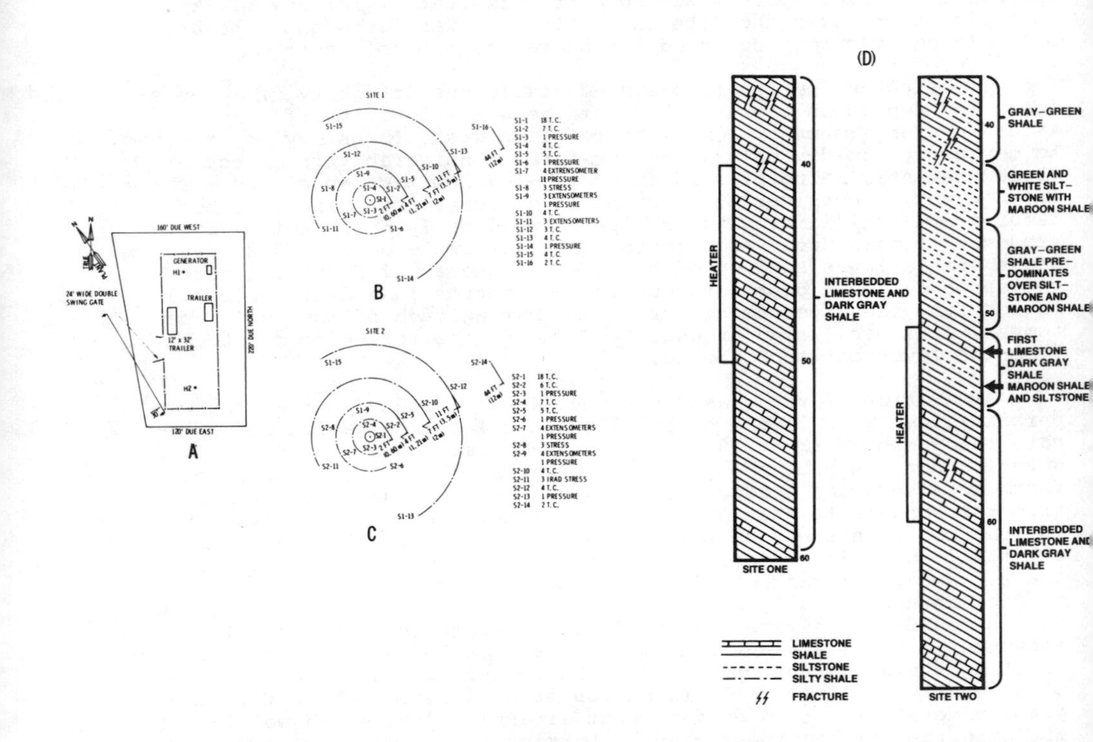

FIGURE 1. (A) Overall Site Layout; (B) Experiment 1 Layout; (C) Experiment 2 Layout; and (D) Lithologies in Heater Holes

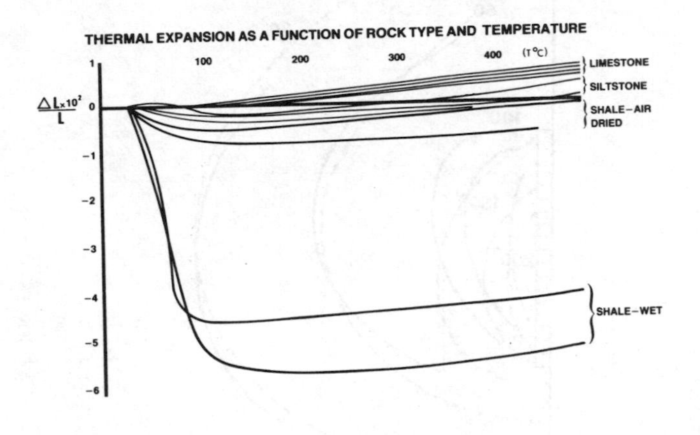

FIGURE 2. Thermal Expansion as a Function of Temperature and Rock Type

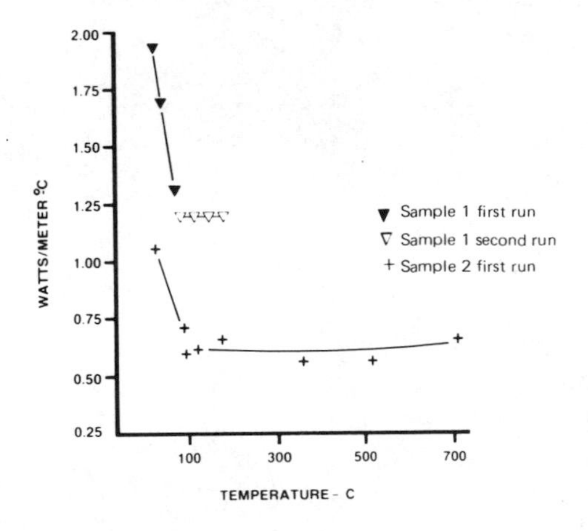

FIGURE 3. Thermal Conductivities of Conasauga Samples

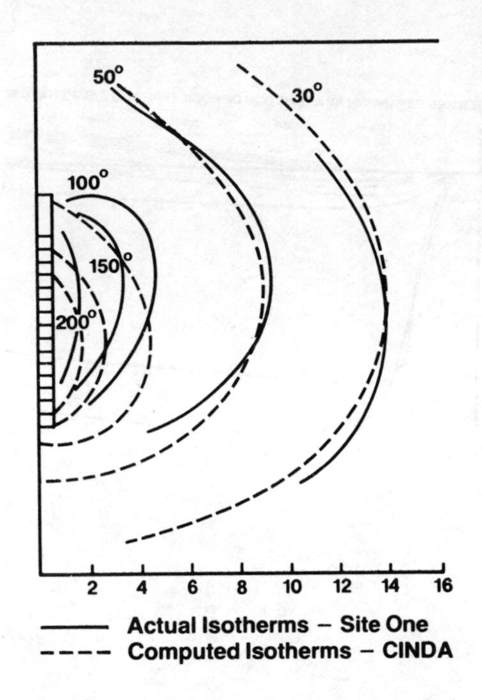

FIGURE 4. A Comparison of Field Isotherms with Computed Isotherms for 135 Days

Discussion

A.A. BONNE, Belgium

Has the condensed steam been analyzed ?

J.L. KRUMHANSL, United States

No, not the condensed steam. The gas phase in the heater hole was tapped as a gas and analyzed. These analyses showed that the composition resembled air except for the fact that it contained water (1 %) and CO_2 (1 %). Neither H_2S or SO_2 were detected during the test.

A.A. BONNE, Belgium

With the results obtained by this experiment, that has run for a period of about one year, would it be possible to make a long-term prediction about the behavior of materials in respect to corrosion ?

J.L. KRUMHANSL, United States

No, I can point out where some problems lie, but since radiolysis products were absent it is not possible to derive corrosion rates that could be used to model the expected canister life under actual repository conditions.

N.A. CHAPMAN, United Kingdom

Referring to the in situ glass corrosion study, I would like to know how firmly the superficial gel was attached to the unaltered substrate. Also was there any variation in thickness of this layer, or was it locally absent ?

J.L. KRUMHANSL, United States

The coating could be easily flaked off using a needle. The layer was not uniform ; it was absent where the sample had been protected by a piece of stainless steel.

R.H. HEREMANS, Belgium

Envisage-t-on ou a-t-on déjà procédé à des mesures de température, perméabilité, etc. "in situ" ou sur échantillons après des temps différents et après arrêt du chauffage. Si oui, la température atteinte était-elle celle du milieu original ?

J.L. KRUMHANSL, United States

Yes, permeability has been assessed after the test while the maximum site temperature was about 25°C ; water samples were taken just prior to and immediately after the test so that the maximum water temperature was about 100°C ; coring was undertaken at a time when the maximum site temperature was about 20°C, and photographs were taken of the heater walls when the temperature was about 20°C. It should also be pointed out that the cooling rate of the site was watched after the heaters were turned off.

F. GERA, NEA

Have you observed any hydrothermal alterations in the shales studied after the experiment ?

J.L. KRUMHANSL, United States

Not of the silicate mineral phases ; we do, however, see oxidation of the pyrite (i.e. $4\ FeS_2 + 8\ CaCO_3 + 15\ O_2 + 6\ H_2O \rightleftarrows 4\ Fe(OH)_3 + 8\ CO_2 + 8\ CaSO_4$) and a bit of oxidation of the Fe^{++} in clays. No change was seen in the X-ray diffraction traces between pre and post test samples, and no new silicate minerals were seen by examining samples taken ofter the test in thin section or with the scanning electron microscope.

F. GERA, NEA

How do you explain the observed decrease of permeability of the formation ?

J.L. KRUMHANSL, United States

There are two possibilities : a) the gypsum formed may have blocked the fractures, which seems unlikely considering the texture of the precipitated gypsum, and b) the site as a whole may have undergone a net expansion, and as a consequence of the confinement provided by the unheated rocks adjacent to the test site, this expansion may have served to close cracks even in regions that had suffered dehydration.

RETENTION DE RADIOELEMENTS A VIE LONGUE PAR
DIVERS MINERAUX ARGILEUX

D. RANCON* , J. ROCHON**

RESUME

La rétention des principaux radioéléments à vie longue contenus dans les déchets a été étudiée sur 8 matériaux naturels : des argiles (Attapulgite, Bentonite, Illite, Kaolinite, Sépiolite, Vermiculite), une zéolite (Clinoptilolite) et la Bauxite. La Kaolinite,moins favorable, exceptée, tous ces matériaux ont des propriétés sorbantes comparables au pH d'équilibre dans une eau faiblement minéralisée : Très forte rétention de Pu, Am, Zr et Sm ; rétention moyenne de Np et Sr ; rétention faible ou nulle des anions iode (I^-) et Tc ($Tc\ O_4^-$). Compte tenu de leurs propriétés physiques, de leur disponibilité et de leur sensibilité aux variations de pH, les matériaux les plus aptes à constituer de bonnes barrières géochimiques seraien la bentonite, l'illite, l'attapulgite et la bauxite.

Le cas important du Pu est à considérer à part, la sorption est très forte mais dans tous les cas, il existe une faible fraction du Pu non retenue et entrainée par l'eau.

ABSTRACT

The retention of the principal radionuclides existing in radioactive wastes has been studied on 8 natural materials : clays (Attapulgite, Bentonite, Illite Kaolinite, Sepiolite, Vermiculite), a zeolite (Clinoptilolite) and the Bauxite. Except for the kaolinite, less favourable, all this materials have comparable sorbing properties at the equilibrium pH in a weakly mineralized water : very strong retention of Pu, Am, Sm, Zr and Cs, middle retention of Np and Sr ; very weak or no retention of the anions iodine (I^-) and Tc($Tc\ O_4^-$). Taking into account of their physical properties, of their availablility and of their sensibility to the pH variations, the more fitted materials to constitute geochimical barriers are the bentonite illite, attapulgite and bauxite.

The important case of Pu must be considered apart, the sorption is very strong but in all the cases, it exists a small fraction of Pu which is not sorbed and which is carried away by the water.

- otection et Sûreté Nucléaire du CEA
herche géologique et Minière

PRESENTATION

Ce document est un condensé des rapports présentés à la réunion de travail de Bruxelles / 1 / et au Congrès d'Otaniemi /2/ aux mois de Janvier et Juillet 1979. Dans le cadre des recherches coordonnées par la Commission des Communautés Européennes[xxx]*, sur les possibilités de stockages de déchets radioactifs dans les structures cristallines profondes, nous avons entrepris une étude sur la qualification de barrières géochimiques, matériaux destinés à faire écran entre le produit stocké et la roche encaissante pour empêcher ou limiter la migration d'éléments radioactifs sous l'effet de courants d'eau éventuels.*

2 - LES RADIOELEMENTS CONSIDERES -

En raison de la longue durée des expériences on a été conduit à ne considérer que les radioéléments de longue période dont la teneur initiale est suffisamment importante pour constituer un danger à long terme après 100 ans et 1000 ans de stockage. Ils sont présentés dans le tableau 1.

Après 1000 ans de stockage subsisteraient en quantité notable les transuraniens et parmi les P.F. compte tenu des rendements de fission essentiellement le ^{93}Zr et le ^{99}Tc.

La production d'^{129}I est d'une part très faible (environ 10^{-7} fois l'activité totale des déchets après 10 ans de décroissance) et d'autre part improbable dans des déchets tels que produits vitrifiés. On l'a toutefois considéré en raison de son danger biologique et pour le cas ou il serait stocké à part.

---

[x] Institut de Protection et Sûreté Nucléaire du CEA

[xx] Bureau de recherche Géologique et Minière

[xxx] Contrat de la Communauté Economique Européenne pour l'Energie Atomique CEE 019-76-7-WASF

TABLEAU 1

LES RADIOELEMENTS ETUDIES

Catégorie	Elément	Isotopes à vie longue (T an)	Isotope utilisé (T)
Transuraniens	Np	237 239	237 ($2,2.10^6$)
	Pu	239 240 242	239 ($2,4.10^4$)
	Am	241 243	241 (470)
Produits de fissions cationiques	Sr	90 (28)	85 (64j)
	Zr	93 ($9,5.10^5$)	95 (65j)
	Cs	137 (30) 135 ($2,9.10^6$)	134 (2a)
	Sm	151 (90)	153 (47h)
P.F. anioniques	Tc	99 ($2,1.10^5$)	96 (4,3j)
	I	129 ($1,6.10^7$)	131 (8j)

3 - LES MATERIAUX ETUDIES -

Après une première élimination de matériaux trop peu efficaces ou trop rares, on a retenu 8 silicoaluminates dont 7 argiles qui sont présentés sur le tableau 2.

TABLEAU 2

LES SILICO-ALUMINATES ETUDIES

NOM	NATURE
Attapulgite	Argile fibreuse
Bauxite	Mélange d'hydrates (Al,Fe) et d'argiles
Bentonite	Argile de la famille Montmorillonite
Clinoptilolite	Zéolite naturelle
Illite	Argile phylliteuse
Kaolinite	Argile phylliteuse
Sépiolite	Argile fibreuse
Vermiculite	Argile Phylliteuse

4 - LA RETENTION -

La rétention a été évaluée par mesure du coefficient de distribution du radioélément entre la solution et la phase solide ; le Kd se prête bien à des mesures comparatives dans des conditions physico-chimiques données comme c'est le cas dans cette étude de qualification et sélection de matériaux sorbants. Par contre le Kd n'est pas toujours suffisant pour les calculs de transferts de radioéléments en milieux poreux (voir /2/). Comme la rétention est très sensible aux variations de pH, on a dans tous les cas mesuré les variations du Kd en fonction du pH. Les radioéléments ont été introduits en trace dans une eau faiblement minéralisée (eau du granite).

Les résultats détaillés de ces expériences sont exposés dans les documents /1/ et /2/.

On peut séparer les radioéléments en quatre catégories :

- les corps peu ou pas retenus : I, Tc
- les corps moyennement retenus par sorption ionique : Sr, Np
- les corps fortement retenus par sorption ionique : Cs
- les corps fortement retenus par précipitation et sorption ionique : Zr, Sm, Pu, Am.

I et Tc se trouvent en solution sous forme anionique I^- (iodure) et TcO_4^- (pertechnétate) ; le k_d décroît quand le pH augmente, la sorption étant d'autant plus faible que la densité de charges positives du minéral argileux décroît.

Le strontium en solution, sous forme Sr^{++} est retenu par adsorption et échange d'ions ; l'augmentation du K_d avec le pH est due à l'augmentation des sites présentant une charge négative.

Le neptunium existe en solution sous forme NpO_2^+ les processus de rétention par les argiles ne sont pas encore connus et il est probable que l'échange d'ions y joue aussi un rôle.

Le césium (Cs^+ en solution) est aussi retenu par échange d'ions, mais de façon plus importante, du fait de son faible rayon ionique hydraté.

La rétention des actinides est liée aux phénomènes de précipitation. Toutefois, la précipitation n'est pas seule en cause, si on compare, à même pH les valeurs des k_d sur les argiles à ceux mesurés sur le quartz où le phénomène de précipitation est prépondérant /1 et 2/. On peut penser qu'avec les argiles, il y a superposition de trois mécanismes : précipitation, sorption des colloïdes rétention par échange du Pu et de l'Am en solution.

La précipitation joue aussi un rôle prépondérant dans la rétention du Sm et du Zr, mais les grandes variations du k_d en fonction du pH ne peuvent recevoir qu'une explication partielle à partir de la distribution théorique des espèces.

Sélection des matériaux sorbants -

La kaolinite exceptée, dont le pouvoir sorbant est nettement moindre pour tous les éléments, les matériaux considérés à des degrés divers selon les radioéléments pourraient constituer de bonnes barrières géochimiques car ils retiennent bien les cations et les transuraniens.

Les critères de classifications des meilleurs matériaux ont été détaillés dans /1/. Compte tenu de leurs propriétés sorbantes, de leur sensibilité aux variations de pH, de leur disponibilité, de leurs conditions de gisement et de leurs propriétés physiques, les matériaux argileux les plus aptes à consituer de bonnes barrières géochimiques sont dans l'ordre.

La bentonite
L'illite
L'attapulgite

Cas particulier du plutonium -

Des essais en colonnes ont été effectuées sur des mélanges argile-quartz à la proportion 98-2 (99,5 - 0,5 pour la bentonite) par la méthode injection-impulsion /2/.

Dans tous les cas, on a remarqué qu'il existait une faible proportion de Pu (0,3 à 0,8 % selon les argiles) entraînée par l'eau et à la même vitesse que l'eau, une autre fraction est lentement entraînée par l'eau, le reste du Pu (96 à 99%) restant fixé de façon apparamment irréversible dans la phase solide.

Des études sont en cours pour déterminer la nature de cette "forme mobile" du Pu qu'on a retrouvé dans d'autres expériences sur des sols naturels, des schistes et des grès.

C O N C L U S I O N

Les matériaux argileux qui par leurs propriétés physiques (conditionnement, étanchéité) se prêtent bien aux diverses opérations concernant la sûreté des stockages, constituent aussi de très bons matériaux sorbant vis à vis de la majorité des radioéléments présents dans les déchets, les plus efficiaces étant la bentonite, l'illite et l'attapulgite. Toutefois ces matériaux ne retiennent pas les anions (iode, technétium) qui peuvent être retenus par des minéraux de toute autre nature (minerais de Cu, Pb ou Fe).

Les recherches se poursuivent par des études sur l'influence de la température, des produits de corrosion sur la rétention et par des expériences sur grosses colonnes pour évaluer la répartition dans la phase solide des radioéléments fortement retenus.

BIBLIOGRAPHIE

/1/ - D. RANCON - R. ROCHON
RETENTION DES RADIONUCLIDES A VIE LONGUE PAR DIVERS MATERIAUX NATURELS -
Réunion de travail de Bruxelles sur la migration des radionuclides à vie longue dans la géosphère.
Agence pour l' Energie Nucléaire de l'O.C.D.E. - 1979

/2/ - D. RANCON - R. ROCHON
Recherche en laboratoire sur la rétention et le transfert de produits de fission et de transuraniens dans les milieux poreux -
Colloque international sur l'évacuation des déchets radioactifs dans le sol.
Otaniemi, Finlande, 2-6 Juillet 1979
RAPPORT I.A.E.A. S.M - 243/155.

Discussion

R. PUSCH, Sweden

You gave Kd values for bentonite in the pH ranges : $pH < 6$ and $pH > 11$. We rely on bentonite as buffer material in the KBS concepts, but we would be worried if pH were lower than 6 or higher than 11 because montmorillonite is not stable outside the pH range 6-11. How did you manage to carry out your experiments with a preserved crystal state outside the range I mention. An X-ray diffraction study would have revealed that the material was no longer montmorillonite.

D. RANCON, France

Il s'agit d'un effet global. La diminution du Kd aux faibles pH peut aussi provenir d'une modification ou d'une dégradation de l'argile surtout aux $pH < 3$. La sépiolite est détruite, en dessous de pH 5 à 6. Les autres argiles se sont révélées plus résistantes.

L.R. DOLE, United States

In the conduct of your experiments, was there an attempt to control the oxygen or the electrochemical potential ?

D. RANCON, France

Ces contrôles ont été effectués uniquement dans les expériences sur le technétium. Les solutions aux divers pH sont restées en conditions oxydantes.

B.R. ERDAL, United States

How was the pH controlled in these studies ?

D. RANCON, France

En général, l'introduction d'argile dans une eau à pH donné, modifie le pH du mélange (pH d'équilibre argile-eau). Aussi, pour étudier la rétention à un pH donné, il est nécessaire de contrôler l'évolution en cours d'expérience et éventuellement d'ajuster le pH à la valeur désirée par adjonction de base ou d'acide.

A. BRONDI, Italy

Vous avez étudié la capacité de plusieurs matériaux argileux de constituer une barrière géochimique supplémentaire aux barrières artificielles. Ces matériaux sont également employés comme matériaux de remplissage. Or, la dimension de la cavité dans laquelle on va placer les canisters est bien délimitée. Des modifications minéralogiques et physiques des argiles utilisées comme barrière pour la chaleur émise par les canisters peuvent se vérifier. Dans ce cas, on peut aussi penser à des modifications plus ou moins importantes des propriétés de la barrière géochimique originale des matériaux argileux. Alors, il faudrait être sûr que ces modifications ne soient pas étendues jusqu'à la paroi granitique du dépôt pour ne pas détruire la barrière géochimique ainsi que sa fonction.

D. RANCON, France

C'est vrai, les propriétés sorbantes des argiles pourront être modifiées par divers facteurs, notamment par l'échauffement et par les produits de corrosion des canisters. Nous avons commencé par le plus simple : la qualification d'une argile non perturbée. Actuellement, on étudie l'influence de l'échauffement sur les trois argiles sélectionnées (bentonite, illite, attapulgile). L'étude de l'influence des produits de corrosion va commencer.

A. AVOGADRO, CEC

A propos des expériences de percolation en colonne avec les transuraniens, avez-vous constaté une variation de la fraction mobile en fonction de la variation de la composition de l'eau, en particulier pour ce qui concerne la concentration des anions complexants ?

D. RANCON, France

Nous avons toujours utilisé le même type d'eau (eau des nappes du granite).

A.A. BONNE, Belgium

Bien que le sujet de cette réunion de travail concerne les argiles, j'aimerais savoir si vous avez considéré également parmi les matériaux de tampons disponibles, les oxydes et hydroxydes de manganèse. Ces substances ont prouvé a plusieurs occasions qu'elles possèdent une capacité d'adsorption non négligeable.

D. RANCON, France

Nous avons utilisé des minerais de Fe, Cu et Pb pour étudier la rétention des anions, mais pas le manganèse. Nous notons cependant votre suggestion.

A.A. BONNE, Belgium

Est-ce que l'eau que vous avez utilisé comme solution mère est une eau présente dans les granites ?

D. RANCON, France

Nous avons reconstitué en laboratoire une eau présentant les caractéristiques moyennes des eaux des granites.

A.A. BONNE, Belgium

Quel est le niveau de concentration des éléments dans la solution pour laquelle vous avez fait les expériences de Kd ?

D. RANCON, France

Nous avons toujours utilisé des radioéléments en trace comme ils devraient se trouver dans les eaux issues des déchets stockés.

THE KBS CONCEPTS — GENERAL OUTLINE, PRESENT STUDY

R. Pusch
Div. Soil Mechanics
University of Luleå
Luleå
Sweden

The Swedish KBS 2 concept, which concerns spent, unreprocessed reactor fuel, implies the use of an "engineered" barrier of highly compacted Na bentonite for isolating metal canisters with the wastes from the surrounding rock. The isolating power of a barrier of this kind will be so great that it will probably be suggested for other radioactive wastes as well.

Fig. 1 illustrates the KBS 2 concept. The copper canister C is isolated from the surrounding rock by Na bentonite. The bentonite is applied in the form of prefabricated blocks obtained by

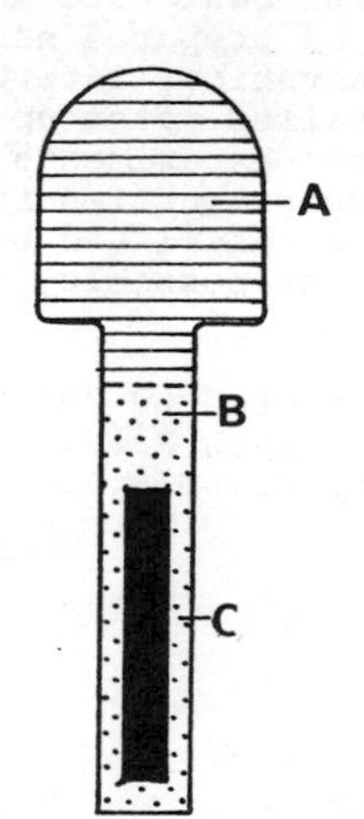

Fig. 1 Cross section through tunnel and deposition hole with canister C containing the radioactive material. A illustrates sand/bentonite backfill compacted in-situ. B represents highly compacted bentonite.

isostatic compaction of granulated bentonite powder using pressures of the order of 50-100 MPa. This gives the bentonite a bulk density of about 2.1 t/m^3 and a degree of water saturation of about 65% when the powder has a water content of about 10% by weight.

After the application of the blocks, water will be absorbed from the rock and the clay will swell and become homogeneous. The coefficient of permeability will be of the order of 10^{-13} - 10^{-14} m/s which means that the extremely low hydraulic gradients (10^{-2} to 10^{-3}), which will exist some hundred years after the closing of a repository, will hardly produce any water percolation at all through the clay barrier. Diffusion will be the only ion transport mechanism that has to be considered. Experiments have shown that the diffusion coefficient is about 1% of that of water. The thermal conductivity of the highly compacted bentonite will be of the order of 0.75 W/m, °C which yields a canister surface temperature of less than 70° C.

The water uptake is produced by the suction potential of the clay, this potential being closely related to the swelling pressure. This pressure is then of great interest since it determines the rate of water uptake. It also affects the rate of bentonite extrusion through joints in the surrounding rock which may be widened as a consequence of an altered stress situation due to tectonics etc.

The rate and uniformity of the water uptake in the deposition holes with special reference to the thermal gradients produced by the warm canisters is of great practical interest. It is being studied at present in "bore holes" consisting of steel cylinders, 1.5 m high and 0.3 m in diameter in the laboratory. A large number of thermo-couples and moisture gauges (resistivity measurement using AC technique) are embedded in the bentonite which also contains a heater so that the thermal situation in a deposition hole is simulated. Very valuable information concerning monitoring and recording techniques have been gained and the experiments have yielded a large number of interesting data.

Parallell to this study, systematic investigations of the swelling pressure, diffusion properties, and permeability are made with reference to the influence of pore water electrolytes, bentonite composition and temperature.

In the Stripa mine, tests are now being prepared for the study of water uptake also in larger bore holes with bentonite and heaters. A series of well instrumented holes will be prepared next year. The holes will be opened and investigated at various intervals over a period of 3-5 years or more so that the detailed water uptake pattern can be found with great accuracy. Tracer experiments will be made for the determination of the diffusion properties for some relevant ion species. A tunnel section with backfill and bore holes will be made as well to simulate, as closely as possible, the natural conditions in a repository.

Tests at higher temperatures than the expected 60-70° C will probably also be made, possibly in cooperation with Prof. Witherspoon's group from Lawrence Berkely Laboratories, U.S.A.

Discussion

D.F. McVEY, United States

Please discuss the time required for the clay (bentonite) to swell and close up the assembly gaps between waste canister and clay blocks and between clay blocks and granite. If these gaps are very large, and require a long time to close, there may be a significant effect on the canister temperatures because of the high thermal resistance of the gaps.

R. PUSCH, Sweden

The gaps are not empty but filled with somewhat compacted bentonite powder. It is true that the compact bentonite blocks will swell but this will lead to a compaction of the bentonite powder in the gaps so that there will never be a state where swelling bentonite expands into an empty space. It is merely a question of a successive internal adjustment of densities which will finally lead to an equilibrium where the bentonite fill is isotropic and homogeneous. The required time could be some decades to some hundred years, or possibly even thousand years.

A.L. NOLD, Switzerland

Could you please give more details about the planned use of prefabricated buffer material blocks ?

R. PUSCH, Sweden

Bentonite powder of the Na Wyoming, granulated type (MX80) with a natural water content of about 10 % will be "isostatically" compacted under a pressure of about 50 MPa which yields a product with a bulk density of 2.1 t/m^3 and a degree of water saturation of about 65 %. The blocks have the appearance of talc and do not dry out or swell in ordinary room climate. They have strength similar to that of fairly soft sedimentary rock and can therefore easily be handled and transported.

L.B. LEWIS, United Kingdom

What are the dimensions in the hole, for example the gap between the canister and the wall ?

R. PUSCH, Sweden

The diameter of the deposition hole is 1.5 m and its depth about 7.7 m. The prefabricated blocks will be somewhat smaller so that they can be placed in the hole without difficulty and so that the Ø 0.77 m canister can be lowered into the central space formed by the ring-shaped blocks. The gaps between canister and blocks, and between blocks and rock will be about 30 mm. They will be filled by bentonite powder which will be applied and somewhat compacted by some suitable ring-shaped device.

GENERAL DISCUSSION DISCUSSION GÉNÉRALE

R.H. HEREMANS, Belgium

Après la présentation suédoise sur les concepts KBS, il serait intéressant de savoir ce qui est envisagé dans les autres pays concernant les expériences sur les matériaux de remplissage.

T.F. LOMENICK, United States

I suppose your question is in regard to tests for backfill materials. At this time I am not aware of any special tests being done for backfilling around containers that might be disposed of in salt or any other medium. Perhaps some of the other people from the US might be more familiar with this subject.

A.R. LAPPIN, United States

All I know is that as far as the experiment in the Eleana argillite is concerned a primary consideration was retrievability. We considered the idea of either packing the heater in place or grouting it in the hole, but decided not to do either one, because we wanted to be able to get the heater back out and examine both heater and emplacement hole. As part of that work we did not do anything about backfilling.

S. GONZALES, United States

It has only been during the last year that a decision to consider multiple barriers was made in our programme.

R.L. DOLE, United States

I think at this point in the US we are trying to organise a programme to look at the corrosion of canister materials in the presence of radiolysis products. I know Sandia has done a preliminary study in this area and some corrosion tests under radiation have been performed. The backfill concept up till last year or so has been to use crushed salt and there has been some work on the consolidation of salt with different moisture contents and the effect of infiltration of water into this salt backfill, but that is the extent of it so far. There was some preliminary work done at Argonne National Laboratory on the use of various minerals as backfill for example pyrite has been considered, but we do not have an extensive programme on the development of backfill materials at this time.

T.F. LOMENICK, United States

This is perhaps premature but there is a very interesting proposal which will be discussed at Oak Ridge in the next few weeks. This has to do with a hot cell test whereby a canister of waste or, perhaps even better today, a spent fuel element would be encapsulated and placed in a simulated disposal environment. One would end up with a spent fuel element surrounded by a mixture of calcium oxide and sand within a block of rock salt. The idea would be to determine the important variables and the effects that, for example, brine radiolysis, would have on the canister or on the spent fuel element. The experiment would be highly instrumented and one might be able to inject brine or any other fluid which might be important. This would provide information on the interactions between calcium oxide, the fill material and the salt. This test is now in its infancy ; the proposal has not yet been officially presented to the Office of Nuclear Waste Isolation and I mention it here because this meeting might be interested in our plans.

R.H. HEREMANS, Belgium

D'autres commentaires sur le programme américain ?

B.R. ERDAL, United States

There are several experiments, concerning radionuclides migration and the chemical properties of various clays and backfill materials that are being undertaken by various US organisations. Some of these tests are similar to those described by Mr. Rançon.

R.H. HEREMANS, Belgium

Puis-je demander aux représentants de la Suisse si ils ont un programme concernant les matériaux de remplissage, et plus particulièrement les matériaux argileux.

A.L. NOLD, Switzerland

In Switzerland, the main efforts of research have been concentrated on investigations of anhydride and crystalline rocks. As regards crystalline rocks the bedrock of the Swiss plain with a waste repository at a depth of 1000 to 2500 m is envisaged. Argillaceous formations have not been given first priority. The interest into argillaceous material is explained by the problem of buffer material for surrounding high-level waste canister. It is planned to join our Swedish colleagues and cooperate with them in the tests to be started next year and stretching over a period of 4 years at the Stripa mine. An intensive drilling campaign is going to be started by the end of 1980 to give more accurate information on the dimensions of possible host rock formations.

R.H. HEREMANS, Belgium

Merci à la Suisse. La Grande-Bretagne.

A.G. DUNCAN, United Kingdom

Mr. Chairman, I am under the impression that at the moment we are speaking about materials to be used as buffer and for backfilling and plugging. In the UK we have been impressed with the work that has been done in the US for the Office of Nuclear Waste Isolation and I am surprised that our US friends have not mentioned it. The report that has been published indicates that there is a lot of information available on backfilling of mines and shaft sealing. The shaft sealing problem is one that we are not really concerned about here, and is similar to the problem of plugging experimental boreholes in the vicinity of a potential repository site. Therefore let me concentrate on our views on the backfilling of a waste repository. The report that I mentioned above points out that there is a lot of information available but that we do not actually know what the backfilling requirements are. What is the function of a backfill? Is it to prevent water from moving about around the waste ? Is it to enhance the adsorption capacity in the vicinity of the waste ? Or is it to control the chemistry of the environment in order to enhance the canister's corrosion resistance or to change the radionuclides chemical forms in order to enhance their retention by the geologic materials ? A number of these questions have to be answered so that we know what are the functions of the buffer material. Then of course we have to answer the question about how long do we expect it to carry out those functions. In the UK we are now at the stage of attempting to formulate a programme and to answer these questions. I would like to ask Dr. Lewis to describe the thinking to date.

J.B. LEWIS, United Kingdom

We are trying to look at the whole system which includes : leaching of materials from the glass, the effect of the corroded canister (because water would only get in contact with the glass if the canister had been corroded), the chemistry of the system as a result of leaching, corrosion of the canister and backfilling with various materials and then the interactions of certain chemical species with the surrounding rock. Since the surrounding geologic materials could be shales or crystalline rocks, and the backfilling material can change the chemical form of elements such as technetium and iodine, selected minerals could be added to change the valence state and achieve better retention. There are many variables and what we are presently doing is trying to see how to optimize the performance of the multiple barrier system. In particular we would like to look at the chemistry of actinides since several of our colleagues have told us that a fraction of plutonium will migrate and what migrates might be much more important than what is left behind. We would like to perform a sensitivity analysis of the migration of actinides in order to identify what fraction of actinides could migrate without a significant impact on the safety of the system. Experimental work is being planned to investigate the chemical and geochemical aspects of the problem, but no results are yet available.

N.A. CHAPMAN, United Kingdom

I think that the borehole plugging techniques which will be applied will probably not involve the use of clay minerals and, therefore, are not really relevant to the topic of this discussion.

A. BRONDI, Italy

Nous n'utilisons pas les matériaux argileux en tant que matériaux de remplissage des cavités dans lesquelles seront placés les canisters, car nous utilisons directement la formation argileuse comme barrière géochimique.

Jusqu'ici nous avons étudié la capacité de capturer des ions dans des zéolites, mais dans notre programme futur, nous prévoyons d'étudier les capacités des argiles de notre territoire national comme barrières géochimiques et de sélectionner celles qui sont les plus adaptées parmi les différentes formations argileuses, ayant différents caractères minéralogiques et géochimiques. Nous avons également entrepris une étude sur les pouvoirs de corrosion des différents types d'argiles qui sont répartis dans le territoire. Puisque nous pensons enfouir directement des déchets dans les argiles, il n'est pas nécessaire d'ajouter d'autres argiles comme matériaux de remplissage.

Dans une situation comme celle qui a été illustrée par nos collègues suédois, et qui n'est pas tellement différente des choix que l'on fait en France, on voit que la dimension de la barrière constituée par l'argile, par rapport à la dimension de la cavité, est assez petite et que les températures seront probablement au-dessous de 100°C. Je pense pourtant que ces températures seront capables de modifier d'une certaine façon la minéralogie ou les propriétés physico-chimiques des argiles et également de modifier l'efficacité de ces barrières artificielles supplémentaires. Je vois alors les argiles de remplissage de la cavité, non pas comme une barrière définitive, mais comme une barrière retardant la dispersion possible des radionucléides. En Italie nous ferons des études générales sur les propriétés géochimiques des argiles et nous utiliserons les résultats qui sont donnés par nos collègues étrangers pour sélectionner les formations dans lesquelles on peut essayer de faire des dépôts.

R.H. HEREMANS, Belgium

Est-ce que du côté de la France, quelqu'un désire ajouter quelque chose ?

D. RANCON, France

En réalité, en France nous ne sommes pas encore très avancés dans ce domaine. Pour le moment, nous étudions les argiles comme barrière géochimique, la corrosion et les effets de la température.

R.H. HEREMANS, Belgium

En Belgique, comme nos collègues italiens, nous envisageons bien sûr de réutiliser l'argile qui aura été extraite lors de la création de cavités souterraines pour le remblayage. Nous n'avons rien fait jusqu'à présent en ce qui concerne la mise en oeuvre de cette argile pour le remblayage, donc technologiquement le problème n'est pas résolu. Nous pensons démarrer à ce sujet dans les prochains mois un programme qui consisterait à ajouter à l'argile naturelle certains additifs ayant une absorption préférentielle pour certains radionucléides comme par exemple le technétium ou le neptunium. Nous suivons donc de très près les études qui sont faites notamment en France sur différents matériaux et il est possible qu'un mélange, disons d'argile naturelle et d'additifs extérieurs serait utilisé chez nous.

K. LUMIAHO, Finland

I should say that I do not know so much about what will be done in Finland because the work is going ahead through the Technical Research Centre. My guess is that we will cooperate with the Swedish in the framework of the Stripa project. We have been doing some geochemical studies, for example Kd values have been measured at Helsinki University. Several clay materials have been studied.

Session 4

Chairman-Président
Dr. N.A. CHAPMAN
(United Kingdom)

Séance 4

A MATHEMATICAL MODEL FOR CLAY BEHAVIOUR IN NUCLEAR-WASTE STORAGE PROBLEMS

D Dirmikis, J Marti, T Maini
Dames & Moore, London, England

C Louis
Simecsol, Paris, France

This paper presents a computer program which has been developed by the Advanced Technology Group of Dames & Moore for solving fully coupled thermo-mechanical problems, in particular, those associated with the disposal of radioactive wastes in clay formations. The program is two-dimensional, Lagragian, and utilises finite-differences in both time and space. Arbitrarily complex constitutive laws and non-linear heat conduction can be easily accommodated due to the explicit formulation of the governing equations. The program can be used in the solution of both static and dynamic problems and shows its best advantages when the thermal and mechanical processes are strongly coupled.

1.0 INTRODUCTION

In recent years there has been considerable interest in the feasibility of solidified radioactive-waste disposal by geological burial. Among the possible host media being currently investigated for this purpose, clays and argillaceous rocks are receiving a great deal of attention in view of their very low permeability and hence high resistance to the transportation of leaking radioactive products by groundwater flow.

However, any practical radioactive waste repository is likely to be subject to very stringent requirements regarding the stability and accessibility of the underground openings during its working life, as well as its long term performance as an efficient barrier to radionuclide migration. In view of the high thermal loads it will have to withstand the required long period of stability, and the novelty of the concept, the design of such a structure cannot be based on existing experience alone. For this reason, and because of the complexity of the equations governing the behaviour of the repository, numerical models have become indispensable for both short and long term predictions of repository performance.

In using such models for thermomechanical analyses of this type of problem, it is necessary to have reasonably accurate data on the relevant physical properties of the host medium surrounding the repository over a wide range of temperatures. For argillaceous materials, a number of experimental in-situ investigations have been reported recently, e.g. Heremans et al [1], Tyler et al [2], aimed at obtaining values for some of these properties. Since high temperatures are likely to develop near the emplaced waste canisters, a computer model has to allow for the possibility of yield and creep of the host medium in the vicinity of the canisters. This requires the development of rather complex rheological models for any realistic representation of clay behaviour. As this paper demonstrates, use of such models requires knowledge of the magnitude and variation of several mechanical properties which can only be obtained by experiment. Thus, in addition to being valuable analytical and predictive tools, computer models also serve as an aid to experimentalists, by highlighting the kind of measurements required for a better understanding of clay behaviour under thermal loads.

2.0 DIFFERENTIAL AND CONSTITUTIVE EQUATIONS

A computer model recently developed by Dames & Moore is based on a two-dimensional, time-dependent, Lagragian formulation of the continuity, momentum and constitutive equations for clay soil elements, coupled to the energy continuity equation for heat conduction through the soil. For two-dimensional problems possessing rotational symmetry, the elements can be considered as toroids with a constant quadrilateral cross-section. In the rest of this treatment rotational symmetry will be assumed.

Accordingly, if r, z and Θ denote the radial, axial and azimuthal directions respectively, the continuity equation for an element is

$$\frac{\dot{V}}{V} = \frac{\partial \dot{r}}{\partial r} + \frac{\partial \dot{z}}{\partial z} + \frac{\dot{r}}{r} \qquad \ldots (1)$$

A general form of the momentum equations that allows mass-proportional Rayleigh damping to be introduced into the system is

$$\rho(\eta\dot{r} + \ddot{r}) = \frac{\partial \sigma_r}{\partial r} + \frac{\partial \tau_{rz}}{\partial z} + \frac{\sigma_r - \sigma_\Theta}{r}$$

$$\rho(\eta\dot{z} + \ddot{z}) = \frac{\partial \sigma_z}{\partial z} + \frac{\partial \tau_{rz}}{\partial r} - \rho g + \frac{\tau_{rz}}{r} \qquad \ldots (2)$$

where

ρ is the soil mass density
$\dot{r},\dot{z}$ are the velocity components in each coordinate direction
$\ddot{r},\ddot{z}$ are the acceleration components in each direction
$\sigma_r,\sigma_z,\sigma_\Theta$ are the normal stress components in each direction
τ_{rz} is the shear stress in the plane of analysis (rz plane)
g is the acceleration of gravity, acting in the negative z-direction
η is the damping coefficient, defined as

$$\eta = \frac{2\pi R}{T_{min}} \qquad \text{... (3)}$$

where R is the fraction of critical damping at the Rayleigh minimum and T_{min} the period at that minimum.
V is the relative volume, defined as

$$V = \frac{\rho_o}{\rho} \qquad \text{... (4)}$$

where ρ_o is the reference (initial) density of the element

The clay constitutive-behaviour model used incorporates a temperature-dependent Von-Mises yield criterion based on a maximum octahedral shear stress

$$f = (\tau_{oct})_{max} = Ae^{\frac{H}{KT}} \qquad \text{... (5)}$$

where

A is a proportionality constant
H is an activation energy
K is Boltzman's constant
and T is the absolute temperature

Below the yield surface defined by (5) the clay behaves elastically and stress increments are defined from strain increments using the theory of thermoelasticity for isotropic materials, i.e.

$$d\sigma_r = (\lambda + 2G_o)\, d\varepsilon_r + \lambda(d\varepsilon_z + d\varepsilon_\Theta) - \beta dT$$

$$d\sigma_z = (\lambda + 2G_o)\, d\varepsilon_z + \lambda(d\varepsilon_r + d\varepsilon_\Theta) - \beta dT$$

$$d\sigma_\Theta = (\lambda + 2G_o)\, d\varepsilon_\Theta + \lambda(d\varepsilon_r + d\varepsilon_z) - \beta dT$$

$$dT_{rz} = G_o\, d\gamma_{rz} \qquad \text{... (6)}$$

where $d\varepsilon_r,d\varepsilon_z,d\varepsilon_\Theta$ are the normal strain increments in each coordinate direction
$d\gamma_{rz}$ is the shear strain increment in the plane of analysis
λ,G_o (shear modulus) are the two Lamé constants
dT is the temperature increment
β is the thermal stress constant, defined by

$$\beta = (3\lambda + 2G_o)\,\alpha = 3\alpha K \qquad \text{... (7)}$$

where α is the thermal coefficient of linear expansion and K is the bulk modulus of the material. All the physical properties above can be functions of temperature.

The components of the deviatoric stress tensor, s_r, s_z, s_θ, τ_{rz} are defined in the usual way, and the octahedral shear stress from

$$\tau_{oct} = \left[\frac{1}{3} (s_r^2 + s_z^2 + s_\theta^2 + 2\tau_{rz}^2) \right]^{\frac{1}{2}} \qquad \ldots (8)$$

If, at any instant in time, the octahedral shear stress exceeds yield criteria of the type given by eq.(5), the creep elements indicated by the one-dimensional rheological model of Figure 1 become operational. The two elements containing dashpots represent primary and secondary creep; the spring constants K_o,K_I and dashpot constants C_I,C_{II} represent the elastic and viscous properties of the material. The two sliders f_I,f_{II} represent the stress levels at which the onset of creep occurs; although they are not necessarily identical, they have been assumed to be equal during this investigation. For equilibrium, the applied stress σ acts across all three elements in Figure 1.

The constitutive laws for each element are

<u>Element O</u> $\qquad \sigma = K_o \varepsilon_o \qquad \ldots (9)$

<u>Element I</u> $\qquad \sigma = K_I \varepsilon_I + C_I \dot{\varepsilon}_I \quad$ for $\sigma > f_I$

$\qquad \varepsilon_I = 0 \quad$ for $\sigma \leq f_I \qquad \ldots (10)$

<u>Element II</u> $\qquad \sigma = (C_{II} \dot{\varepsilon}_{II})^{1/n} \quad$ for $\sigma > f_{II}$

$\qquad \varepsilon_{II} = 0 \quad$ for $\sigma \leq f_{II} \qquad \ldots (11)$

By definition $\quad \varepsilon = \varepsilon_o + \varepsilon_I + \varepsilon_{II}$

hence $\quad \dot{\varepsilon} = \dot{\varepsilon}_o + \dot{\varepsilon}_I + \dot{\varepsilon}_{II} \qquad \ldots (12)$

From eq.(10) $\quad \dot{\varepsilon}_I = \dfrac{\sigma - K_I \varepsilon_I}{C_I} \qquad \ldots (13)$

Substitution of eqs.(9) and (10) in eq.(12) gives

$$\dot{\sigma} = K_o \left[\dot{\varepsilon} - (\dot{\varepsilon}_I + \frac{\sigma^n}{C_{II}}) \right] \qquad \ldots (14)$$

Generalising to three dimensions by assuming that eqs.(13) and (14) apply to deviatoric stresses and strains s_{ij}, e_{ij}, $(e_{ij})_I$ we have

$$(\dot{\varepsilon}_{ij})_I = \frac{s_{ij} - 2G_I (e_{ij})_I}{C_I} \qquad \ldots (15)$$

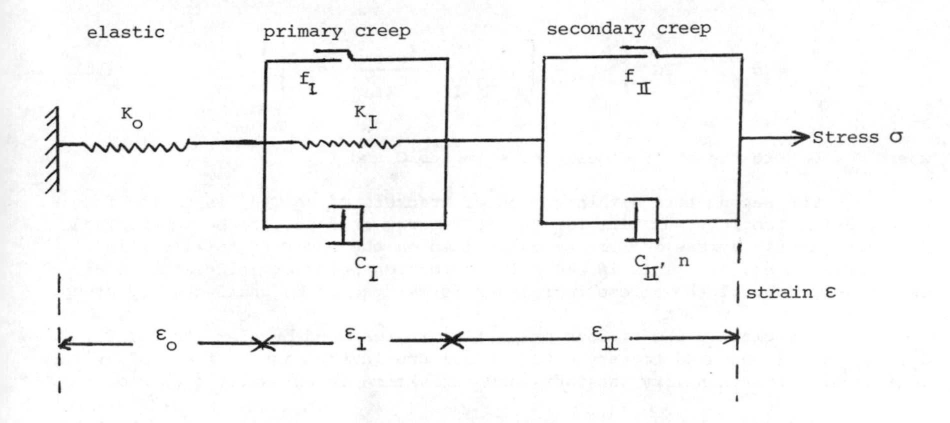

FIGURE 1. RHEOLOGICAL MODEL FOR CLAY

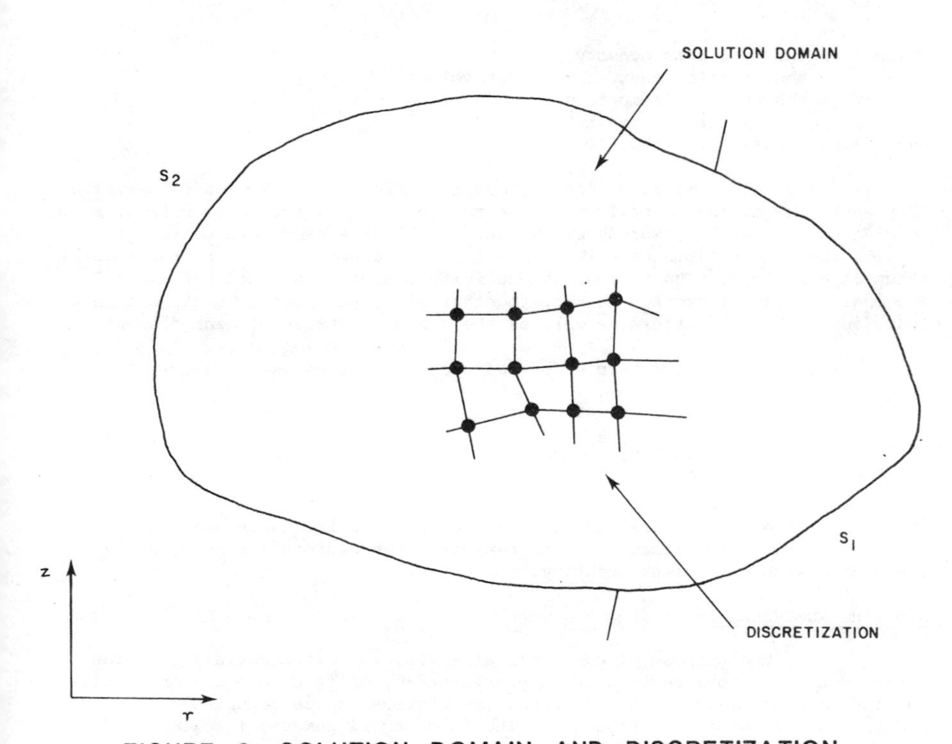

FIGURE 2. SOLUTION DOMAIN AND DISCRETIZATION

$$\dot{s}_{ij} = 2G_o \left\{ \dot{\varepsilon}_{ij} - \left[(\dot{\varepsilon}_{ij})_I + \frac{\tau_{oct}^{n-1}}{C_{II}} s_{ij} \right] \right\} \quad \ldots (16)$$

where G_o, G_I are the shear moduli for elements O and I.

The second term within the outer brackets of eq.(16) is in the form of a plastic "correction" term for the deviatoric stresses, to be subtracted from the elastic stress increments calculated on the basis of total strain increments in eq.(6). This is the only correction necessary since the volumetric component of the stress increments is assumed to be unaffected by creep.

The elastic and viscous properties represented by elements G_O, G_I, C_I, C_{II} are all assumed to vary with temperature in the manner of eq.(5), with individual proportionality constants and activation energies for each property.

Regarding the solution of the energy equation, it is assumed that heat flow within the region of interest takes place by conduction only, which means that the medium considered must either be dry or, if wet, that convection effects are negligible. The power density of any heat sources present can be either constant or varying with time, as would be the case for a radioactive decaying source. In formulating the energy equation, it is further assumed that strain energy changes due to the mechanical deformation process are negligible compared with thermal energy changes due to heat conduction, which is reasonable for the slow deformations considered in this type of problem. Thus, the equation can be written as

$$\rho c_v \frac{\partial T}{\partial t} = \text{div} (\kappa \text{ grad } T) + H \quad \ldots (17)$$

where ρ is the soil mass density
c_v is the specific heat at constant volume
T is the absolute temperature
κ the thermal conductivity
and H the heat-source power density

The values of ρ, c_v and κ above are effective or lumped values, i.e. they apply to all the constituents of a soil element, water, air, solids, etc. considered as a whole. Moreover, the use of a single temperature field T implies that the various constituents within the element are in thermal equilibrium at all times. The thermal conductivity tensor is assumed to have only diagonal non-zero terms κ_r, κ_z but, together with the specific heat, can be a function of the temperature as well as the state of stress of each element.

The heat source term H for radioactive sources is given by

$$H = \sum_{i=1}^{N} H_{o_i} e^{-\lambda_i t} \quad \ldots (18)$$

where H_o is the source strength at time t=0 and λ is the decay constant. The number of terms N is chosen so as to represent the radioactive power decay of the source with sufficient accuracy.

3.0 THE NUMERICAL MODEL

For the purpose of numerical analysis, the differential equations given in the previous section are approximated by difference equations applied to a network of elements that describe the physical space occupied by the medium, as illustrated in Figure 2. The fundamental numerical scheme used involves the solution of the energy equation and the soil deformation equations in an alternating fashion as illustrated schematically in Figure 3.

It can be seen that the solution of the deformation equations will at any given stage proceed to mechanical equilibrium independently of the thermal calculations and vice-versa. However, the two processes still remain coupled through the variation in the mechanical and thermal properties of the medium, e.g. thermal conductivity can decrease due to material decrepitation resulting from large thermal stresses and similarly the shear modulus can decrease as the temperature rises.

The above partially decoupled solution procedure is physically justified in view of the enormous difference in response time (typically about nine orders of magnitude) between the process of soil deformation under the influence of a thermo-mechanical force imbalance and the associated process of heat conduction under the influence of a non-uniform temperature field. Another consequence of this solution scheme is that the soil elements can be treated as either Lagrangian (i.e. fixed mass) elements or Eulerian (i.e. fixed volume) elements, depending on which equations are being solved.

During the heat-conduction calculation stage the elements are treated as normal Eulerian control volumes. This means that their volume, shape and position remain fixed, but heat can flow in and out of them and can also be stored internally. Equation (17) is approximated by a set of first-order, explicit, finite-difference equations applied to the network of elements in Figure 2. Time is subdivided into increments that are small enough to ensure numerical stability. At each time-step, heat fluxes across the four faces of each element and associated incremental changes in internal energy within the element are computed, using finite-difference approximations for the temperature gradient at the mid-point of each element face. When this has been done for all elements, the entire mesh is scanned again in order to update the temperature distribution, and also the distribution of all properties dependent on temperature.

Following this, the mechanical deformation calculations commence. These allow each element to distort under the influence of the thermal stresses set up by the new temperature distribution computed above. The numerical scheme for these calculations is based on a Lagrangian methodology; each element contains a constant mass of material that moves and distorts in space and time, i.e. the element volume and shape may change. Groups of contiguous elements of the same material are described by the constitutive equations given in the previous section which are coupled to the equations of motion. A first-order, explicit formulation of the finite-difference equations is used throughout and time is automatically subdivided into increments that are small enough to guarantee numerical stability.

At the beginning of each time-step the whole mesh is scanned in order to update the velocities and coordinates of the grid points on the basis of a known system of external loads and internal stresses computed from the previous time-step or given as initial conditions. To do this, a contour integral is taken around each grid point in order to derive an equivalent grid-point force from the surrounding stresses. This force is then used to accelerate the grid-point. The acceleration is integrated twice numerically to give the new velocities and grid-point coordinates. When these have been calculated for all grid-points, the complete mesh is scanned again, this time in order to compute incremental strains for each element from the known velocities and coordinates of the four surrounding grid-points. Stresses are then derived from the strains and strain rates using the constitutive laws, i.e. the elastic equations (6) followed, if yielding has occurred, by the appropriate visco-plastic corrections. The latter are derived by simultaneous solution of the system of coupled equations (15) and (16).

The explicit formulation used in this treatment means that the strain state of each element is "frozen" for one time-step, since the time-step is taken so small that information is unable physically to propagate to the next element. This means that the equations of motion for all grid-points are

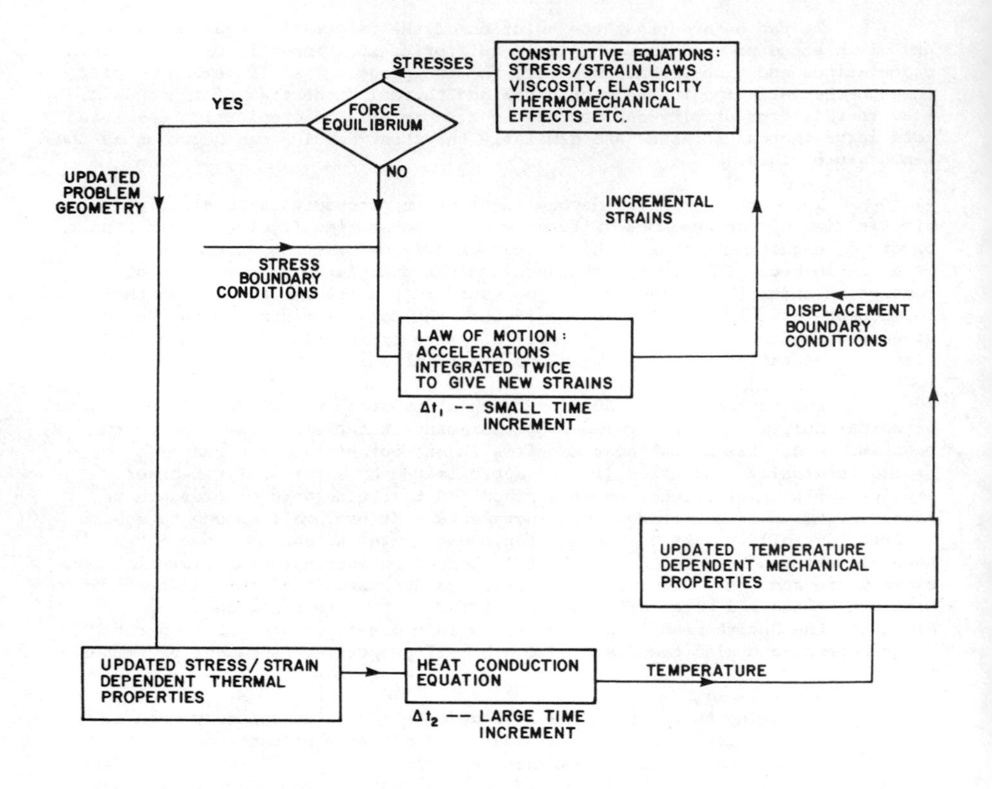

FIGURE 3. ILLUSTRATION OF NUMERICAL SOLUTION SCHEME

physically uncoupled. No iterations are necessary, in contrast to implicit methods where a perturbation in one element will affect all other elements during the time-step.

The mechanical calculations continue until the system reaches force equilibrium. All stress- or strain-dependent thermal properties are then updated and time is advanced by a further increment, during which heat is again allowed to flow, thus setting up a new temperature distribution and commencing a new cycle of the solution algorithm. These cycles can be repeated to cover a specified length of time, or until overall thermo-mechanical equilibrium is reached.

4.0 NUMERICAL RESULTS

The thermal and mechanical calculation components of the solution algorithm were tested independently of each other against several problems for which analytical solutions are known. Two of these are described below.

The first test case involved calculation of the steady-state temperature distribution within an infinitely-long, hollow cylinder whose inner and outer surfaces are maintained at two different temperatures. The analytical solution to this problem (Carslaw and Jaeger, [3]) is

$$T = T_o + \frac{T_i - T_o}{\ln(\frac{b}{a})} \ln (\frac{b}{r})$$

where T_i is the temperature at the inner radius a
T_o is the temperature at the outer radius b
and r is the radial distance from the axis

For the numerical calculations T_i was set to 220°C, T_o to 20°C and $\frac{b}{a}$ to 10. The computed temperatures at various points along the cylinder radius are shown in Figure 4. It is seen that they are in excellent agreement with the theoretical distribution curve also plotted there.

The second test case involved using the code to predict steady-state creep flow for an infinite, thick-walled, hollow cylinder loaded by a uniform internal pressure. This problem possesses an analytical solution (Odqvist [4]), if small strains are assumed, which can be compared with the results of the numerical model. The small-strain approximation was implemented in the code by maintaining a constant geometry rather than updating the nodal coordinates at each time-step. The problem is illustrated in Figure 5 and was chosen to be identical to the one analysed by Anderson [5] so that the results could be directly compared with a finite-element solution. In terms of the properties given in Section 2, the parameters for this problem are G_o=7.692 x 10^6 Pa, K=1.667 x 10^7 Pa, n=4.4 and C_{II}=8.077 x 10^{15} $Pa^{4.4}$ sec. Only secondary creep was allowed, by setting the onset of primary creep, f_I, to a very large value. The fundamental time increment of the numerical process was 0.01 sec.

Figures 6 and 7 show time histories of the radial velocity and the octahedral shear stress at the inner radius of the cylinder and illustrate the attainment of steady-state creep after about 6 secs. The steady-state value of τ_{oct} computed was 356.8 Pa as compared with 364.4 Pa for the finite element code. The corresponding value for the radial velocity was 0.4319 10^{-5} compared with 0.4741 10^{-5} for the other code and 0.4777 10^{-5} for the analytical value. Although still in good agreement with these two values, a somewhat lower accuracy should be expected in our results since, although the same number of elements was used, the finite-difference elements have constant rather than linearly varying strain.

An example of a fully-coupled thermomechanical problem that was analysed is shown in Figure 8. This represents a computer model of a typical near-surface heater experiment in clay, with a 1.5m long, 0.3m diameter heater

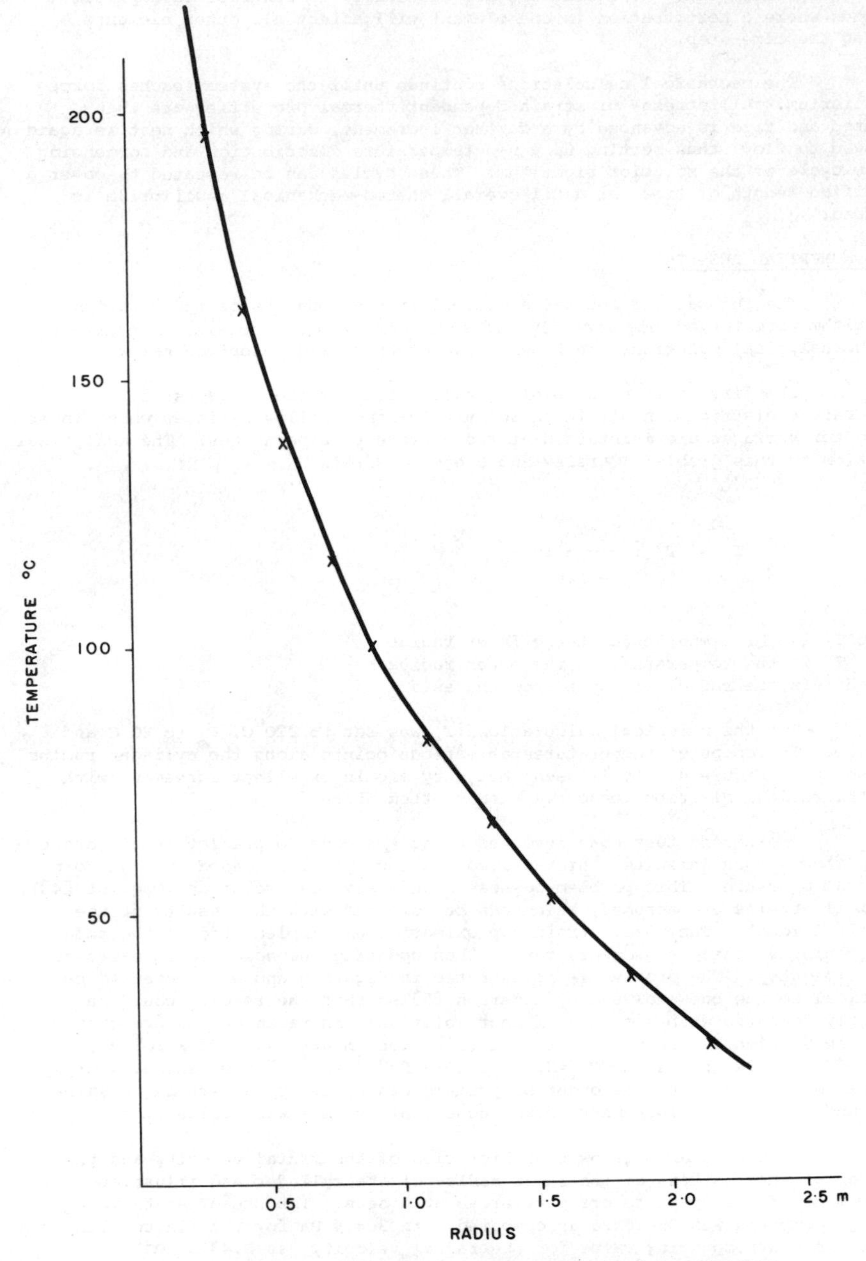

FIGURE 4. STEADY - STATE TEMPERATURE DISTRIBUTION IN HOLLOW CYLINDER

buried 5m below the ground surface and producing a constant 2KW of power. Figure 8 also shows the finite-difference mesh employed and the idealised boundary conditions around the periphery of the solution domain. For the thermal calculations these are a constant temperature of 15°C along the ground-surface boundary AB and adiabatic conditions along the other three boundaries. For the mechanical calculations, the conditions are vertical roller boundaries along AD and BC, a horizontal roller boundary along DC and a free boundary along AB. The in-situ stresses in the soil prior to activation of the heat source were calculated by assuming a vertical lithostatic stress and a coefficient of lateral earth pressure at rest of 0.56, based on a drained Poisson's ratio of 0.36.

The mechanical and thermal properties of clay used are summarised in Table I. For simplicity, and due to the lack of adequate data, the thermal properties were assumed to remain constant throughout the calculation. The mechanical properties that were allowed to vary with temperature are indicated in the table. In each case the functional form of the variation was assumed to be of the type given by eq.(5), the proportionality constants being determined from the values of the properties at room temperature in the table. These values are either typical of argillaceous materials or, in the case of the creep properties, estimates based on little more than intuition, since data on the influence of temperature on clay creep appears to be scarce. In particular, phase changes in the water-clay medium have been disregarded, not because they were deemed unimportant, but because the existing information is not considered sufficient for proposing a sound mathematical description of these thermo-mechanical and chemical processes.

Figure 9 shows the time history of the temperature at a clay element adjacent to the heater and midway along its length, for an elapsed time of approximately 3 days. It will be seen that the temperature rises relatively steeply at first but its rate of increase falls off with time as the temperature gradients become smaller. Figure 10 shows the corresponding time history of the octahedral shear stress at the same location. It can be seen that the stress builds up under the influence of the thermal load and reaches a maximum after about 14 hours. At this point the clay starts to yield, and continues to do so as the yield limit of the material continues to decrease with increasing temperature. The extent of plastification up to that time is illustrated by the shaded elements in Figure 8.

Figure 11 shows isothermal contours 13 days after the heat source was switched on, starting with the 20°C isotherm and proceeding in increments of 20°C. Figure 12 shows the principal stress tensors in the clay at the same time. Figure 13 shows the same tensors for a different analysis, under the conditions of zero gravity and zero in-situ stresses, and illustrates the stresses set up due to the thermal load alone. The calculations were not continued for a larger period of time because the high temperatures developing in the clay and the consequent dewatering imply a drastic change of its properties which is not accounted for in the present model. However, once further experimentation has been carried out to collect the relevant data, the computer model could be easily enhanced to extend its applicability to higher temperature regimes.

5.0 CONCLUSIONS

A numerical model for clay behaviour under thermal and mechanical loads was described which represents a potentially powerful analytical tool for the design of safe nuclear-waste repositories in argillaceous geological media. At the moment, evaluation of the usefulness of this tool is severely restricted by the scarcity of experimental data on the high temperature behaviour of clays, particularly the creep behaviour. However, the existence of such models, apart from their potential long term usefulness, helps to define the kind of further experimentation needed in this field, as well as to help interpret the results of these experiments once they are carried out. The computer program has been constructed so that arbitrarily complex constitutive behaviours can be easily embodied in the program, as justified by new experimental data.

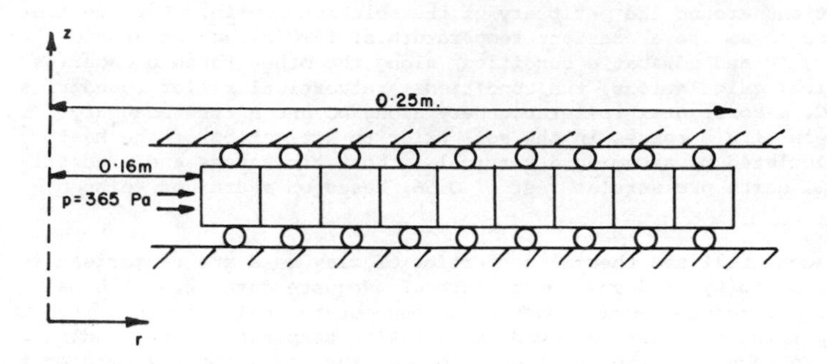

FIGURE 5. THICK - WALLED CYLINDER LOADED BY INTERNAL PRESSURE

REFERENCES

1. Heremans, R, Buyens, M and Manfroy, P: "Le Comportement de l'Argile vis-a-vis de la Chaleur", Proc. of Seminar on In-Situ Heating Experiments in Geological Formations, pp.13-30, Ludvika/Stripa, Sweden, September 13-15, 1978.

2. Tyler, L D, Cuderman, J F, Krumhansl, J L and Lappin, A: "Near-Surface Heater Experiments in Argillaceous Rocks", pp.31-43, Ibid.

3. Carlslaw, H S and Jaeger, J C: "Conduction of Heat in Solids", 2nd Edition, Oxford University Press, Oxford, 1959.

4. Odqvist, F K G: "Mathematical Theory of Creep and Creep Rupture", 2nd Edition, Oxford University Press, Oxford, 1974.

5. Anderson, C A: "Cavity Stability: Finite-Element Analysis for Steady and Transient Creep", Los Alamos Scientific Laboratory, Report LA-5769, February, 1975.

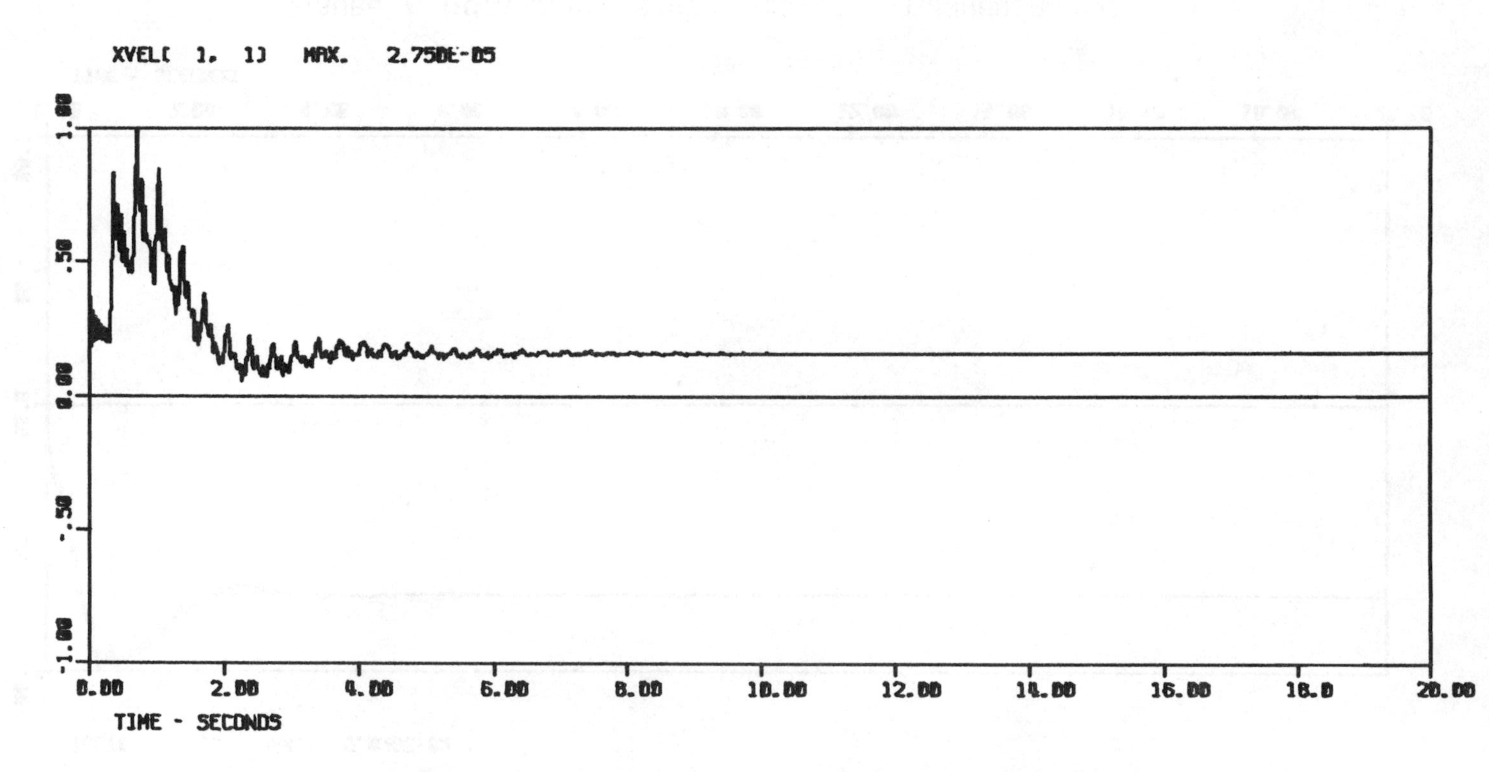

FIGURE 6. RADIAL VELOCITY HISTORY AT INNER RADIUS

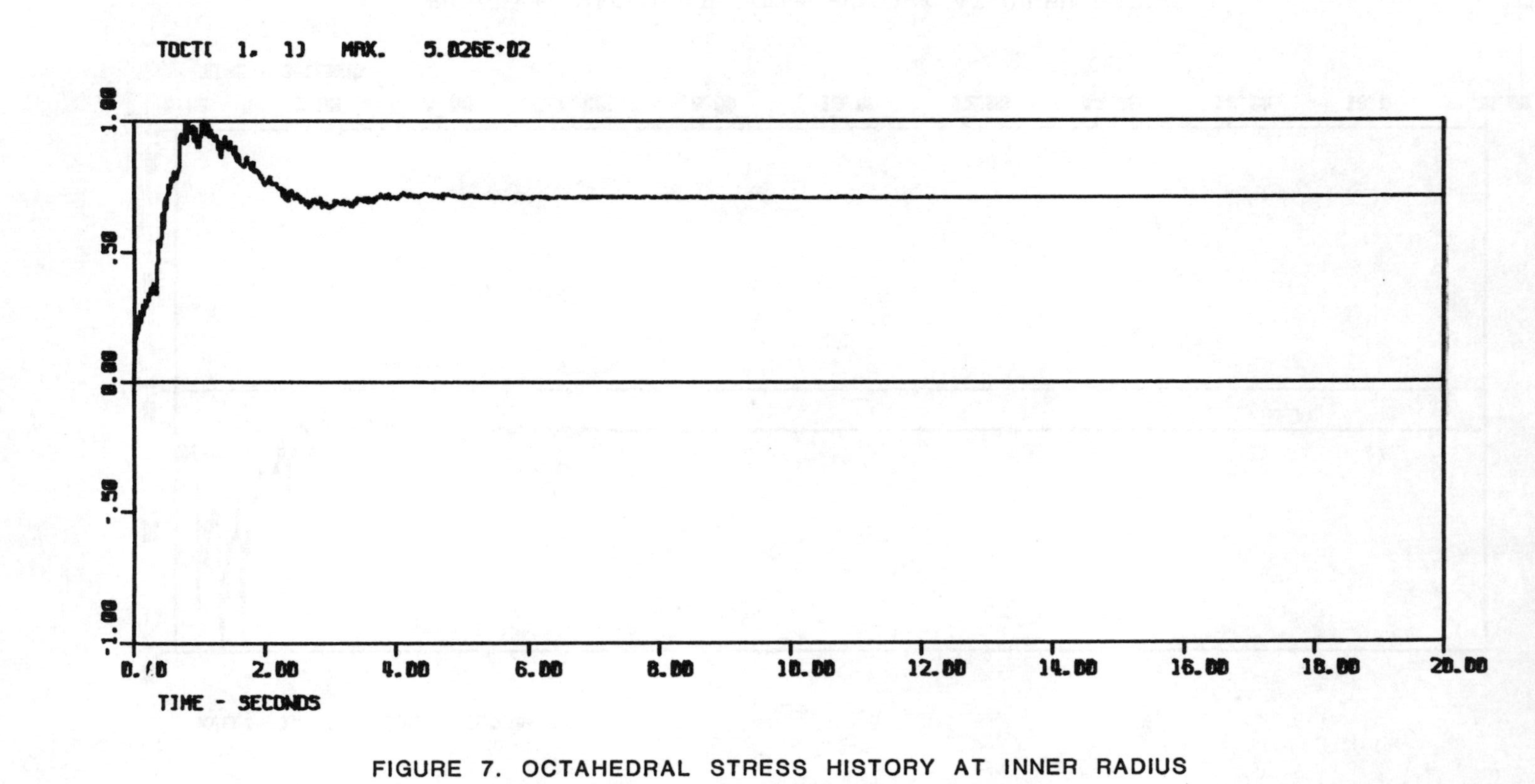

FIGURE 7. OCTAHEDRAL STRESS HISTORY AT INNER RADIUS

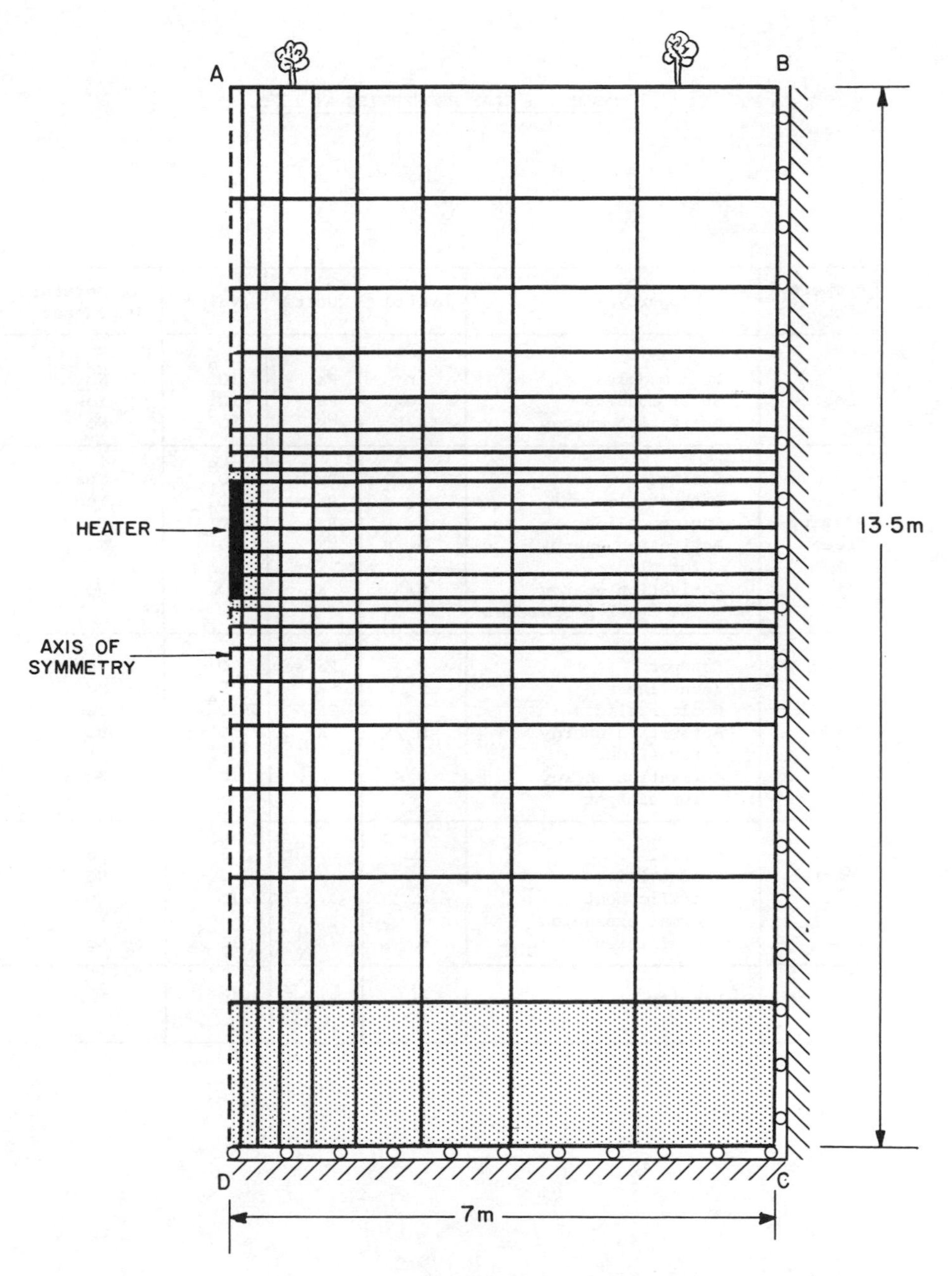

FIGURE 8. FINITE - DIFFERENCE MESH FOR NEAR - SURFACE HEATER EXPERIMENT IN CLAY

TABLE I : CLAY PROPERTIES AT 15°C

Property Group	Property	Symbol	Units	Value	Temperature Dependent
Elastic	Bulk modulus	K	Pa	10^9	No
	Shear modulus	G_o	Pa	10^7	Yes
	Activation energy	H/K	°K	10^3	No
Primary Creep	Spring	G_I	Pa	10^8	Yes
	Dashpot (linear)	C_I	Pa sec	10^{12}	Yes
	Coulomb slider	f_I	Pa	5×10^4	Yes
	Activation energy for slider	H_1/K	°K	2×10^3	No
	Activation energy for other properties	H/K	°K	10^3	No
Secondary Creep	Dashpot (non-linear)	C_{II}	$Pa^n sec$	10^{30}	Yes
		n	-	5	No
	Coulomb slider	f_{II}	Pa	5×10^4	Yes
	Activation energy for slider	H_2/K	°K	2×10^3	No
	Activation energy for dashpot	H/K	°K	10^3	No
Thermal	Thermal Conductivity	κ_r	W/m°K	2	No
		κ_z		2	No
	Specific heat	C_v	J/Kg°K	2×10^3	No
	Thermal expansion coefficient	α	$°K^{-1}$	10^{-5}	No
	Density	ρ	Kg/m^3	2×10^3	Yes

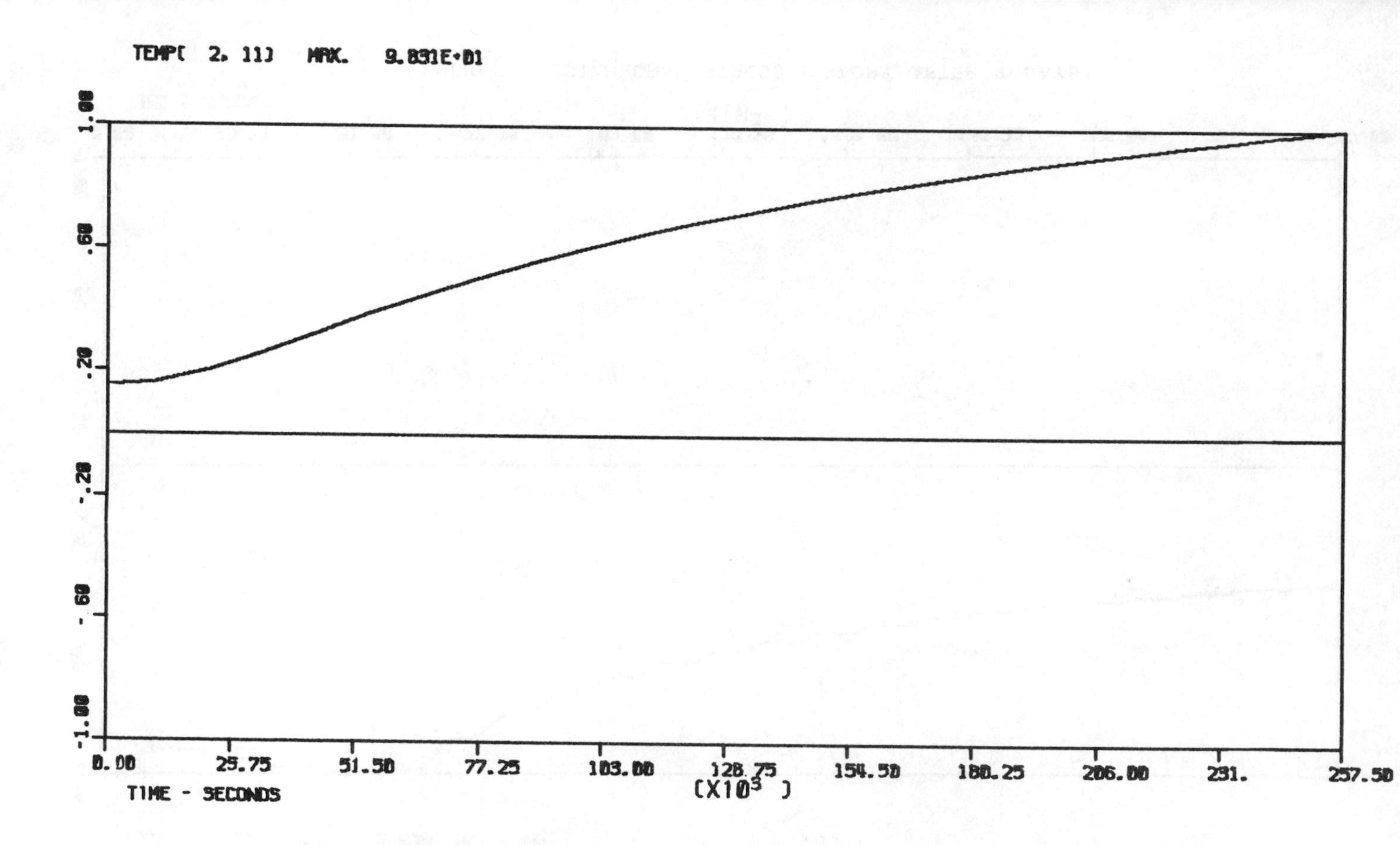

FIGURE 9. TEMPERATURE HISTORY AFTER 3 DAYS

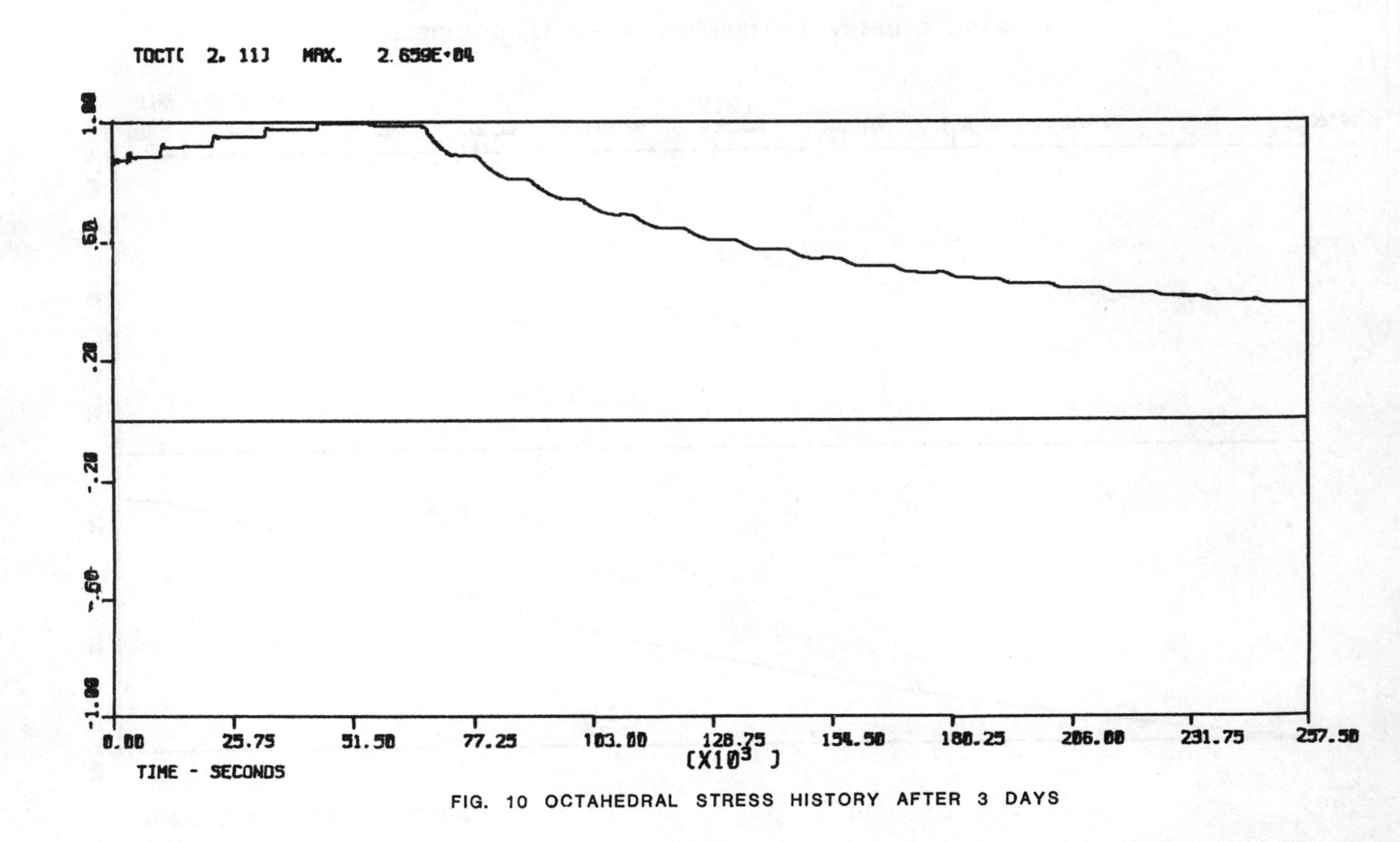

FIG. 10 OCTAHEDRAL STRESS HISTORY AFTER 3 DAYS

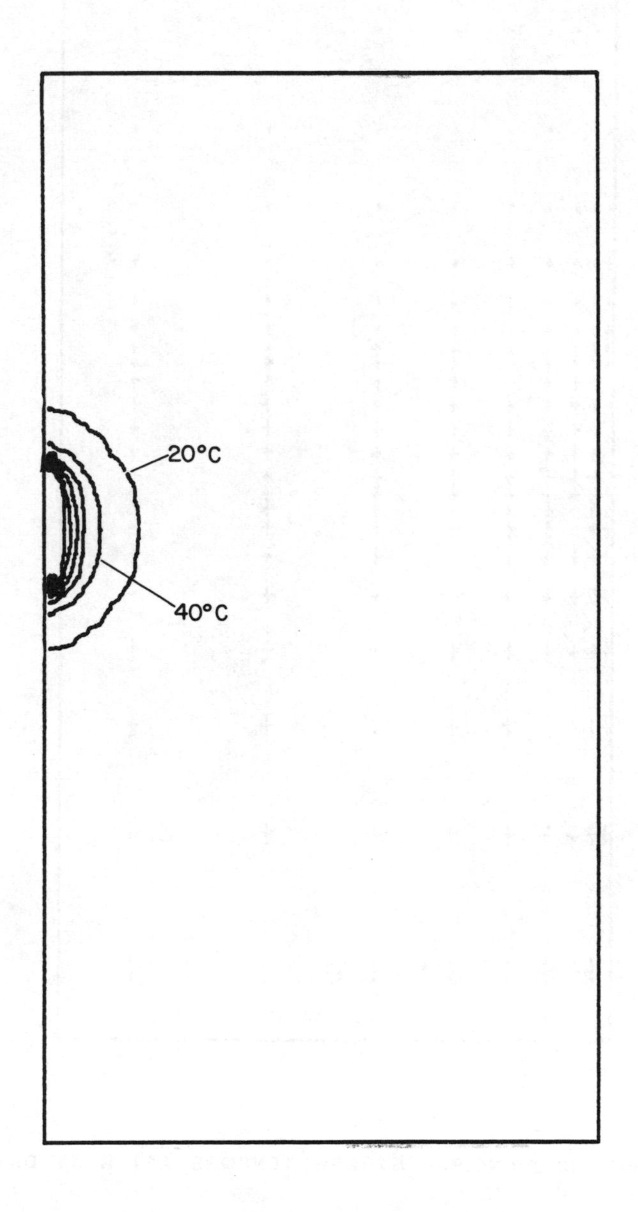

FIGURE 11. TEMPERATURE ISOTHERMS AFTER 13 DAYS

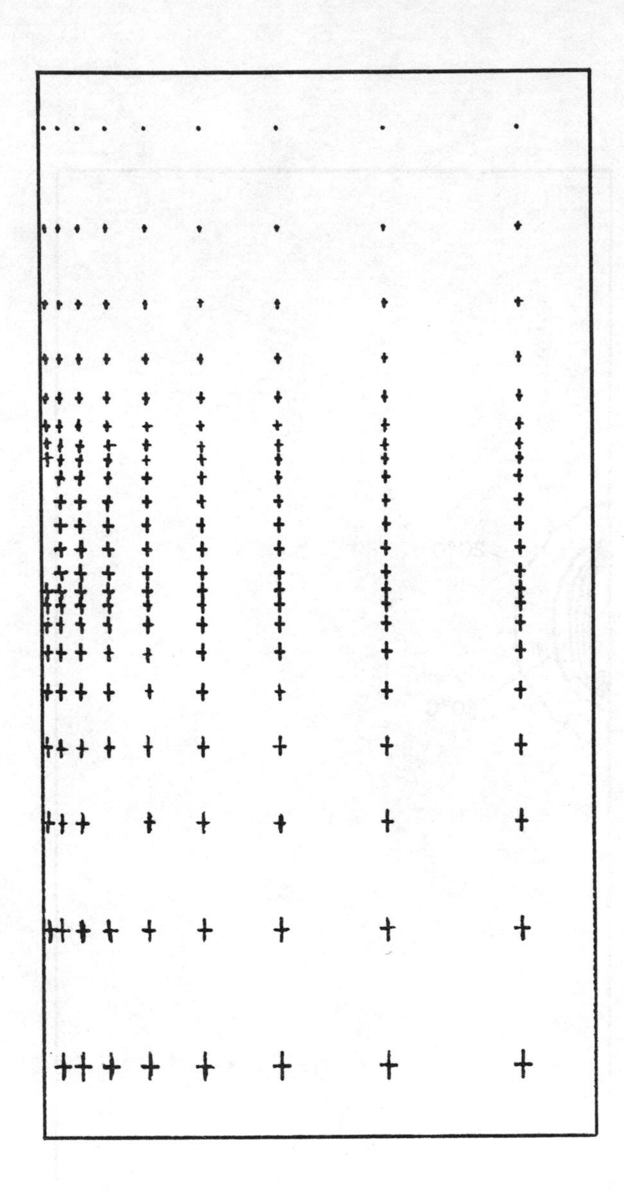

FIG. 12 PRINCIPAL STRESS TENSORS AFTER 13 DAYS

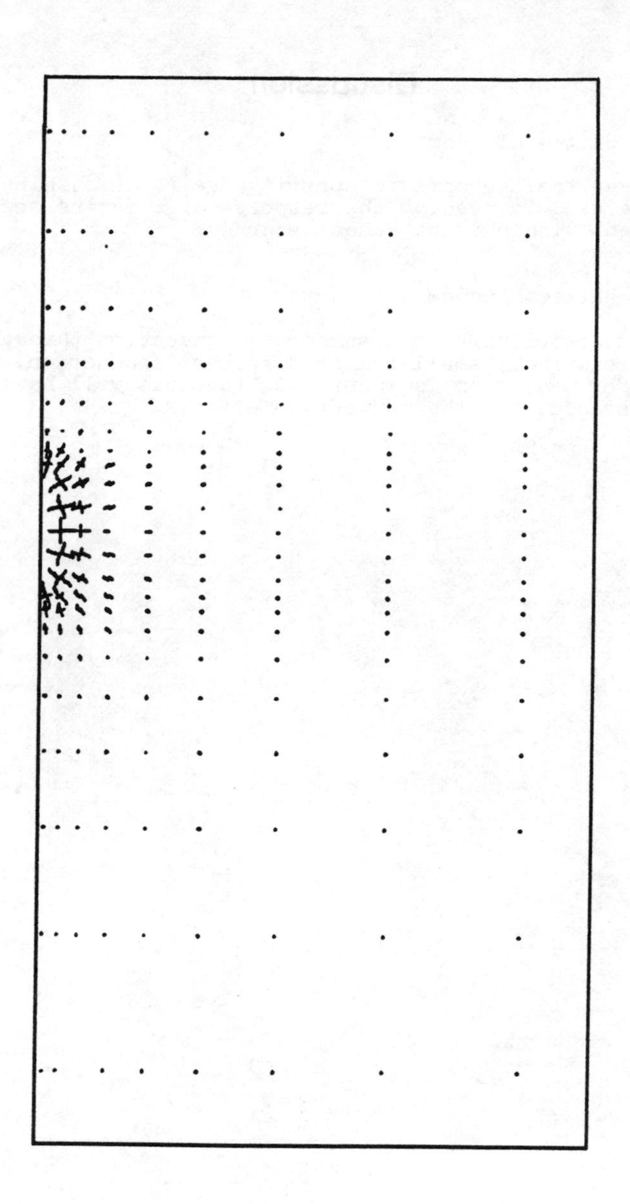

FIGURE 13. PRINCIPAL STRESS TENSORS AFTER 13 DAYS
(THERMAL LOAD ONLY)

Discussion

A.G. DUNCAN, United Kingdom

Given the appropriate constitutive relationships, could this model be used to predict the response of a marine sediment to the emplacement within it of a heat source ?

D. DIRMIKIS, United Kingdom

Yes, provided heat transfer by convection through the sediment is relatively small compared to heat conduction. If the former transfer mechanism is important, the code will have to be upgraded to account for the convective effects.

THERMAL/FLUID MODELING OF THE RESPONSE OF SATURATED MARINE RED CLAYS TO EMPLACEMENT OF NUCLEAR WASTE*

D. F. McVey, D. K. Gartling, and A. J. Russo
Sandia Laboratories**
Albuquerque, New Mexico, U.S.A.

ABSTRACT

In this report, we discuss heat and mass transport in marine red clay sediments being considered as a nuclear waste isolation medium. Development of two computer codes, one to determine temperature and convective velocity fields, the other to analyze the nuclide migration problem, is discussed and preliminary results from the codes reviewed. The calculations indicate that for a maximum allowable sediment/canister temperature range of 200°C to 250°C, the sediment can absorb about 1.5 kW initial power from waste in a 3 m long by 0.3 m diameter canister. The resulting fluid displacement due to convection is found to be small, less than 1 m. The migration of four nuclides, ^{239}Pu, ^{137}Cs, ^{129}I, and ^{99}Tc were computed for a canister buried 30 m deep in 60 m thick sediment. It was found that the ^{239}Pu and ^{137}Cs, which migrate as cations and have relatively high distribution coefficients, are essentially completely contained in the sediment. The anionic species, ^{129}I and ^{99}Tc, which have relatively low distribution coefficients, broke through the sediment in about 5000 years. The resultant peak injection rates which occur at about 15,000 years were extremely small (0.5 µCi/year for ^{129}I and 180 µCi/year for ^{99}Tc).

*This work was supported by the U.S. Department of Energy under contract DE-AC04-76DP00789.

**A U.S. Department of Energy facility.

INTRODUCTION

Participants in the United States Seabed Disposal Program are investigating the technical and environmental feasibility of emplacement of nuclear waste in the deep ocean sediments as a possible disposal option [1]. Of particular interest are illite and smectite clays located in oceanic mid-plate, mid-gyre regions below about 4000 m depth. The clays in the regions of interest are attractive [2,3] in that they have good cation retention characteristics, low permeability, are vertically and laterally uniform over large areas, and are relatively plastic,promoting self-healing. Dating studies of a 24 m core from the Pacific study area show a continuous depositional record for about 76 million years,demonstrating extremely good geologic stability [2,3]. There is nothing at this time to indicate that these clays, in a properly selected oceanic location, do not have the properties necessary for effective isolation of nuclear waste.

For the design of a disposal system, it is necessary to develop the ability to model the interaction of heat and radiation from the waste with the sediment. Specifically, predictions of waste, canister, and sediment temperature histories, sediment pore pressure, pore water motion, thermochemical reactions between seawater/sediment/canister/waste form, sediment structural response, radionuclide migration, and nuclide biological concentration are necessary to define problem areas and propose solutions, develop an optimum design, and assure that the system design is safe. To this end, a modular solution system consisting of a series of computer codes to model important phenomena is being developed. At present, model development efforts embody: thermal energy transport and pore fluid motion, thermochemistry, radiolysis, radionuclide migration, corrosion and leaching, canister and sediment motion, and ocean/biological transport of nuclides.

Code development is supported by active laboratory and field experimental programs in each area [2,3] (e.g., see the results of thermochemical and thermal property experiments reported by Krumhansl and Hadley [4] in a companion paper). The overall program philosophy is to develop a theory and model, obtain required input data from laboratory or field experiments, and evaluate model predictions with field experiments. If required, the theory, model, or input data are refined or modified until an accurate prediction is obtained.

Two of the computer codes under development for seabed disposal modeling will be discussed in this paper. The first of these codes treats the thermal and fluid transport in a porous matrix while the second solves a species transport equation using temperature and velocity fields provided by the first code. These two codes, although still under development, are being used for preliminary studies and it is appropriate to discuss the results to date.

THERMAL AND FLUID TRANSPORT

A transient, non-linear, two-dimensional (planar or axisymmetric) finite element code (MARIAH) incorporating the Darcy equation and Boussinesq approximation has been developed [5,6] to compute the heat and incompressible fluid transport through a rigid, anisotropic porous matrix. Assuming the Boussinesq approximation (1) is valid and dropping the inertial term in the Darcy momentum equation, (2) [7] the equations for an axisymmetric geometry become:

(1) The fluid is incompressible except insofar as the thermal expansion produces a buoyancy force. This assumption is standard in natural convection problems and has been used by many authors in addressing problems of natural convection in porous media.

(2) It has been shown that the error in neglecting the transient phase when using Darcy's law is small since the transition time of the fluid from rest to steady motion is quite small.

Continuity:

$$\frac{1}{r}\frac{\partial}{\partial r}(rV_r) + \frac{\partial V_z}{\partial z} = 0 \qquad (1)$$

Darcy's Law (Momentum):

$$\langle V_r \rangle + \frac{k_{rr}}{\mu}\frac{\partial}{\partial r}\langle P \rangle^* = 0 \qquad (2)$$

$$\langle V_z \rangle + \frac{k_{zz}}{\mu}\left(\frac{\partial \langle P \rangle^*}{\partial z} + \rho_f g\right) = 0$$

Energy:

$$(\rho C)_{eff}\frac{\partial T}{\partial t} + \rho_{f_o} C_f \left(V_r \frac{\partial T}{\partial r} + V_z \frac{\partial T}{\partial z}\right)$$

$$- \frac{1}{r}\frac{\partial}{\partial r}\left[r(\lambda_{rr} - \phi E_{rr})\frac{\partial T}{\partial r} + r(-\phi E_{rz})\frac{\partial T}{\partial z}\right]$$

$$- \frac{\partial}{\partial z}\left[(-\phi E_{zr})\frac{\partial T}{\partial r} + (\lambda_{zz} - \phi E_{zz})\frac{\partial T}{\partial z}\right] - Q = 0 \qquad (3)$$

also,

$$(\rho C)_{eff} = \phi \rho_{f_o} C_f + (1 - \phi)\rho_s C_s \qquad (4)$$

$$\lambda_{ij_{eff}} = \phi \lambda_f + (1 - \phi)\lambda_{s_{ij}} \qquad (5)$$

$$\rho_f = \rho_{f_o}[1 - \beta(T - T_o)] \qquad (6)$$

where,

ρ = fluid density

C = heat capacity

β = bulk expansion coefficient

T = temperature

t = time

$r;z$ = radial; axial coordinates

k_{ij} = permeability tensor

P = pressure

ϕ = porosity

μ = viscosity

V_i = Darcy velocity

E_{ij} = thermal dispersion tensor

λ_{ij} = thermal conductivity tensor

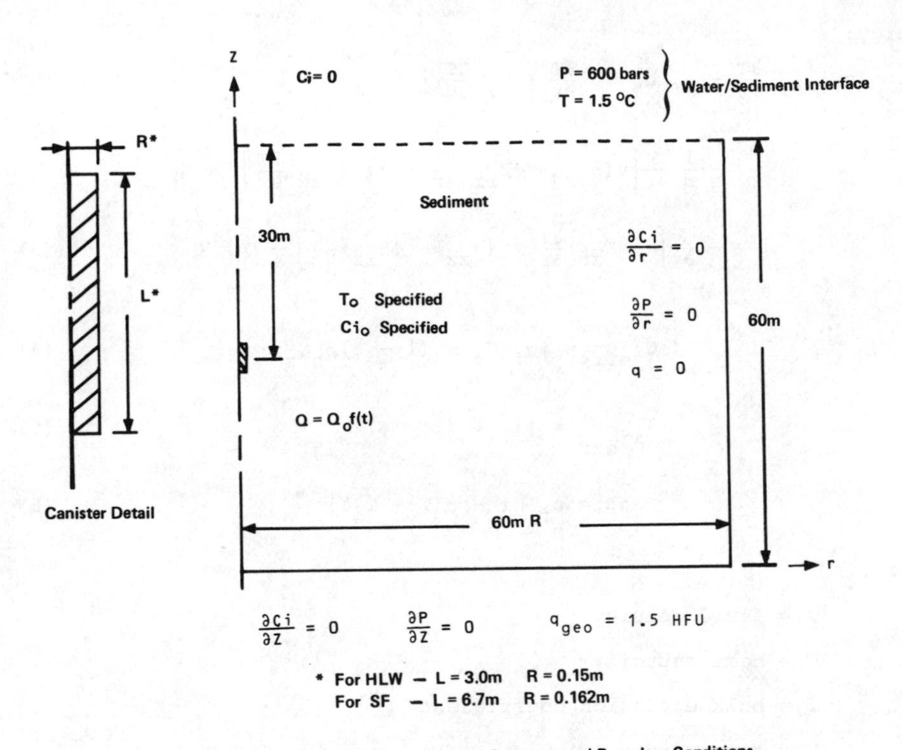

FIGURE 1. Seabed Canister, Model Geometry and Boundary Conditions

$\langle\rangle$ = bulk volume average = $\langle\rangle^* \phi$

$\langle\rangle^*$ = pore volume average

Q = volumetric heat generation rate

Subscripts

f = pore fluid

s = sediment mineral

o = reference value

For solution, Equations 1 and 2 were combined into a Poisson equation for the pressure.

These equations, along with appropriate initial/boundary conditions for pressure and temperature were converted by a method of weighted residuals into a form suitable for finite element application. The details of the code development and users instructions are provided by Gartling and Hickox [5,6].

The MARIAH code has been used to study the heat transfer from waste canisters emplaced in seabed sediment. As a reference problem, the waste is assumed to be emplaced in undisturbed sediment such that the center of the canister is 30 m deep in a 60 m thick layer of sediment. A 1.5 HFU (3) geothermal heat flux was applied to the impermeable lower boundary. The outer boundary at 60 m radius is also impermeable and is assumed insulated. This distance is greater than the thermal penetration in the problem time, about 100 years. The top boundary is at a constant pressure of 600 bars corresponding to a depth of about 6000 m. A constant temperature of 1.5^{o}C equal to the water temperature,is imposed on the upper boundary. The initial temperature distribution in the sediment was determined by the steady state geothermal heat flux and the 1.5^{o}C surface temperature. The geometry along with the boundary conditions is shown in Figure 1.

The canister power output is a function of time due to the waste decay. The normalized power histories for two types of waste for which calculations were made are given in Figure 2. Shown are thermal decay curves for a typical reprocessed high level waste (HLW) and spent fuel (SF). The HLW is assumed to result from a uranium only reprocessing cycle in order to leave the Pu in the waste for the long term heating and nuclide migration problem. The HLW is emplaced 10 years after reprocessing with reprocessing 160 days out of core. The SF is assumed to be a spent reactor fuel assembly emplaced 10 years and 160 days out of core. Isotope inventories for the decay calculation were determined by use of the ORIGEN [8] code.

Material properties used in the analysis are given in Table I. For sediment, the thermal properties are input for the solid and fluid components and the effective properties computed from Equations 4 and 5. The applicability of this approach was confirmed by comparison with the data of Krumhansl and Hadley [4]. Pore fluid properties are for pure water. Appropriate ranges of sediment porosity and permeability were obtained from the data reported by Silva [2,3,9].

The MARIAH code was used with the foregoing input to determine the maximum sediment/canister temperature as a function of initial waste thermal power. The results are summarized by the curve in Figure 3.

(3) Heat flux unit: 1 HFU = 1×10^{-6} cal/cm^{2}s.

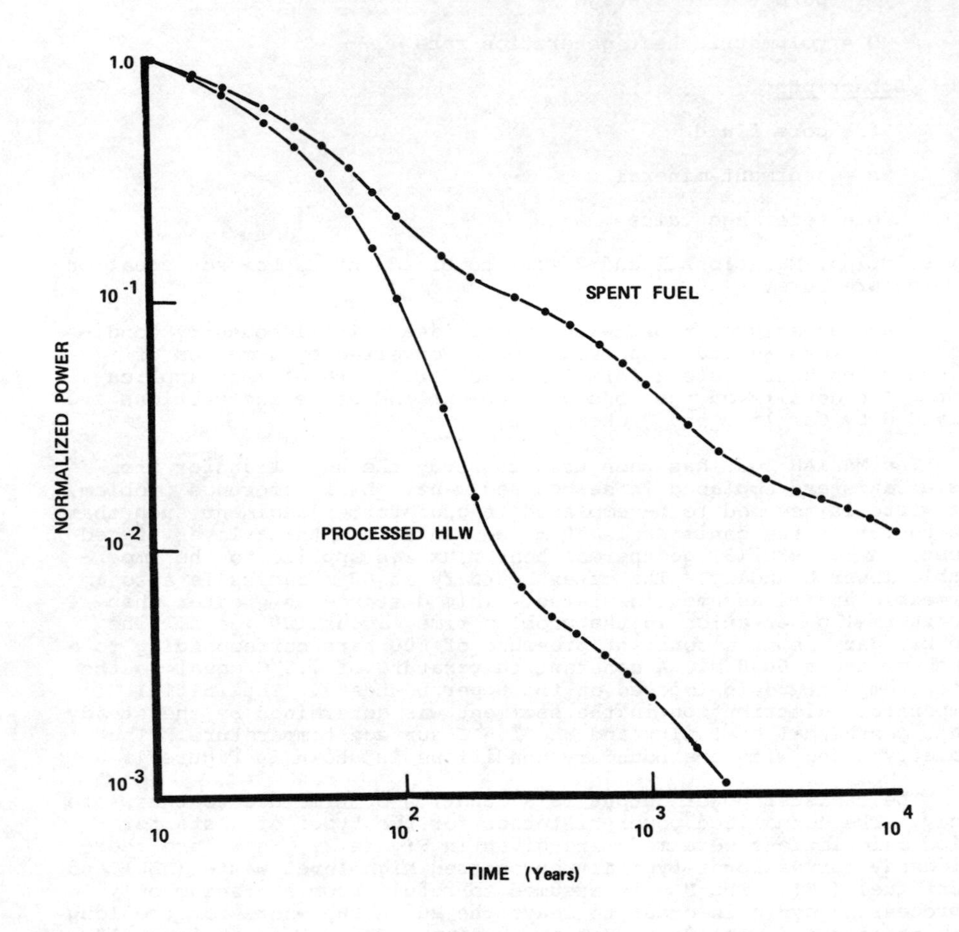

FIGURE 2. Normalized Power Histories for Reference Wastes

TABLE I

Material Properties for Seabed Thermal Analysis - MARIAH

Sediment

	Units	Pore Fluid	Mineral
ρ	kg/m^3	1000.0	2950.0
C	$\frac{W\text{-day}}{kg\text{-}^oC}$	0.05208	0.01018
λ	$\frac{W}{m\text{-}^oC}$	$0.62071 + 0.0013769T - 4.584 \times 10^{-6}T^2 + 3.8889 \times 10^{-10}T^3$	$1.92 \exp(-6.37 \times 10^{-4}T)$
μ	$\frac{kg}{m\text{-day}}$	$1441.15T^{-0.8987}$ $T > 25^oC$ $148.45 - 27.43T$ $T \leq 25^oC$	----
β	$\frac{1}{^oC}$	$0.25 \times 10^{-3} + 4.4 \times 10^{-6}T$	----
ϕ		----	0.8
K_{zz}	m^2	----	5.0×10^{-17}
K_{rr}	m^2	----	5.0×10^{-16}

Waste Solid

	Units	Value
ρ	kg/m^3	2275.0
C	$\frac{W\text{-day}}{kg\text{-}^oC}$	0.010
λ	$\frac{W}{m\text{-}^oC}$	$1.21 + 0.0038T$

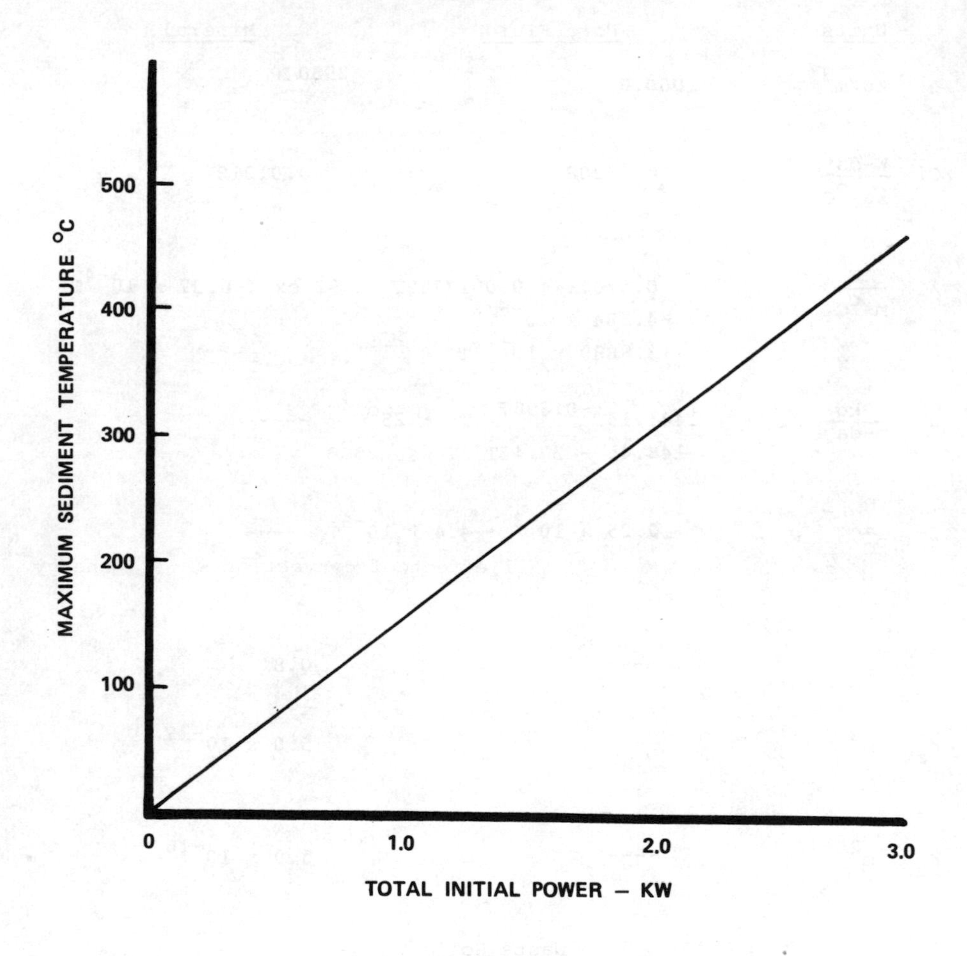

FIGURE 3. **Maximum Sediment Temperature as a Function of Initial Thermal Power**

One of the program design goals is to develop a canister that will contain the waste for several thermal half-lives. From studies of physical and thermochemical changes in sediment and corrosion effects of seawater, sediment, and brines, Krumhansl [4] and Braithwaite [10] recommend limiting the maximum canister temperature to about 200°C to 250°C to meet this goal. For this temperature range, Figure 3 gives a maximum canister power of about 1.5 kW. The curve is applicable for both HLW and SF.

Temperature histories at seven locations radially outward from the canister are given in Figure 4a for 1.5 kW, HLW. Comparable SF temperature curves are shown in Figure 4b for a canistered single Pressurized Water Reactor fuel assembly with 0.58 kW initial power. The surface temperatures peak quite rapidly (in about 1 to 2 years) and then fall off with a half-maximum about equal to the initial waste thermal half-life of about 30 years for HLW and 40 years for SF. The peak temperatures are about 220°C for HLW and a relatively benign 48°C for the canistered SF assembly.

The convective velocity histories for four locations corresponding to temperature locations in Figure 4 are shown for HLW in Figure 5. Since the convection is thermally driven, the velocity curves have essentially the same character as the temperature curves; peak at about 1 to 2 years and decay with a half-maximum about equal to the waste half-life. For HLW, the convective velocity decreased from a maximum of 0.02 m/year to about the same magnitude as the velocity due to molecular diffusion after about 25 years. In the case of SF, of 0.58 kW initial power, the maximum convective velocity is about the same as the velocity due to molecular diffusion. It may be inferred from the velocity history that the total convective contribution to the displacement of the water molecules is small, less than 1 metre. The low velocity is, of course, due to the low permeability of the seabed sediment. Because of the low velocity, the contribution of convection to both energy and nuclide transport is expected to be quite small. The relatively small effect of convection on the energy transport was demonstrated by Hickox [11] who showed analytically that for Rayleigh numbers less than about 0.1, the temperature field is dominated by conduction. In the present case, the Rayleigh number is in the range 10^{-3} to 10^{-4}, depending on temperature.

Isotherm patterns for the 1.5 kW, HLW at 1 and 10 years are shown in Figure 6. It is noted that the maximum extent of the 100°C isotherm, the temperature below which it now appears that significant hydrothermal alteration of the sediment will not occur, is less than about 0.8 metre from the canister. It is also apparent, from Figure 4a, that any significant thermal effects will be negligible in about 100 years, i.e., the canister surface temperature is about 25°C at that time.

Nuclide Migration Calculations

To assess the suitability of the seabed sediment as a nuclear waste containment medium, it is necessary to determine the time required for nuclides that may be released from the canister to reach the sediment surface and to know the quantity and rate of nuclide injection into the seawater. To this end, a radionuclide migration code, IONMIG, is being developed. Details of the code development are discussed by Russo [12]. The two-dimensional planar or axisymmetric code solves a species transport equation including convection, molecular diffusion, axial and transverse dispersion, concentration-dependent sorption, and radioactive decay. In the formulation, it is assumed that the radionuclide concentration remains dilute enough such that fluid properties are not altered and that sorption is non-selective, instantaneous, linear, reversible, and thus describable by an empirically determined equilibrium coefficient which may be a function of concentration. The velocity and temperature fields are input from the MARIAH calculations. The near-field hydrothermal

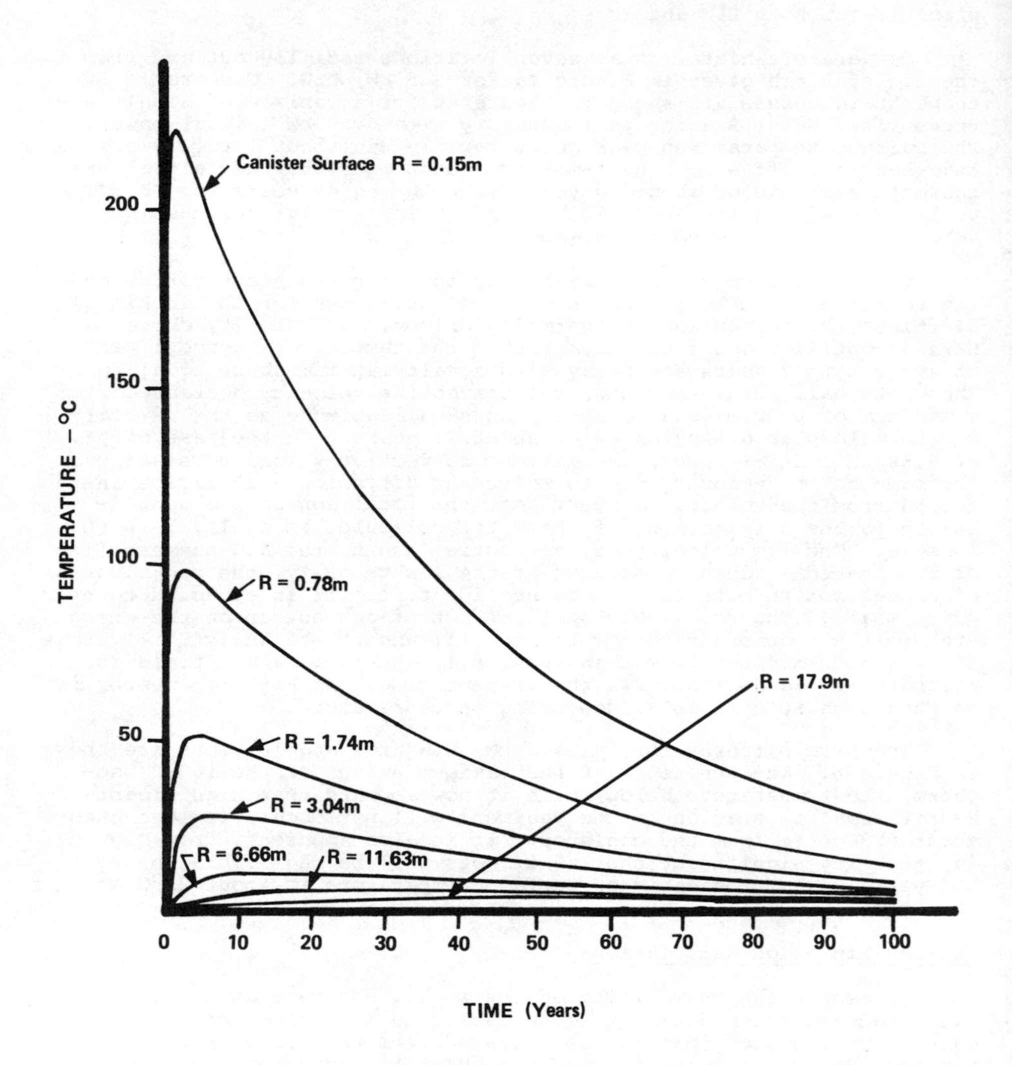

a) HLW, P = 1.5KW

FIGURE 4. Temperature at Seven Locations Radially Outward from Center of Canister as a Function of Time

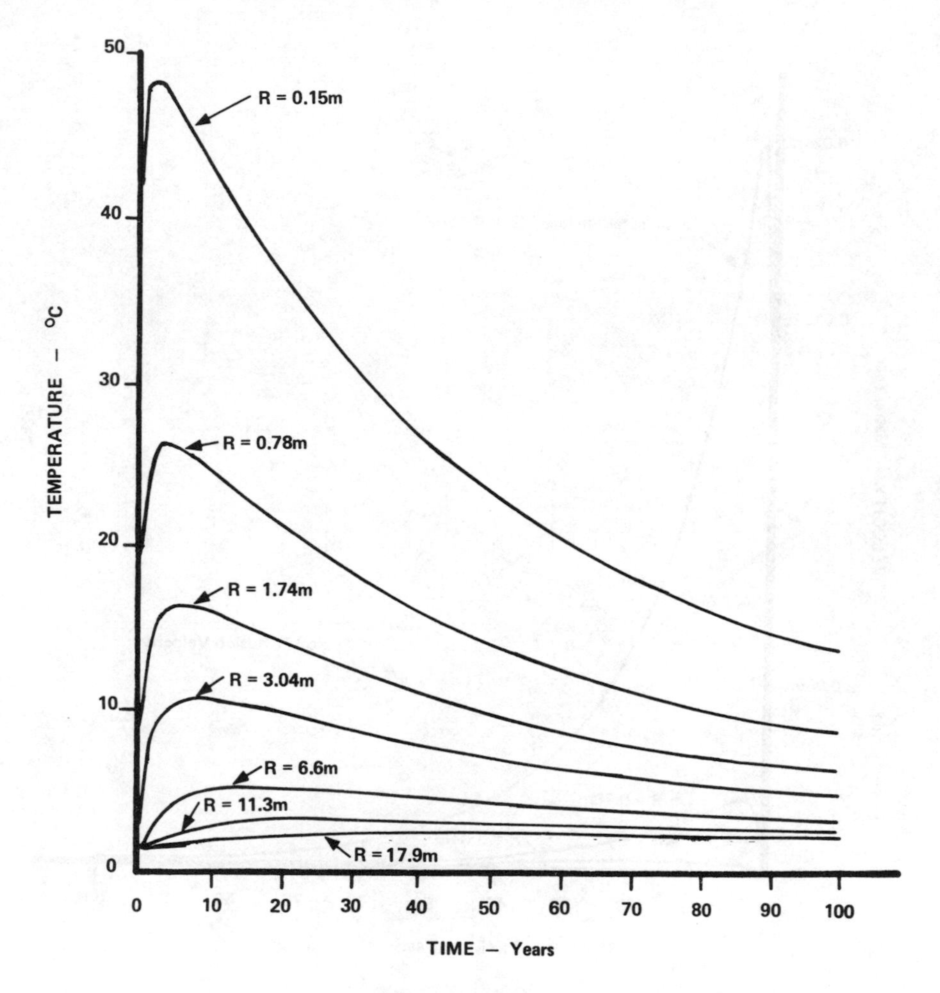

b) SF, P = 0.58KW

FIGURE 4. Temperature at Seven Locations Radially Outward from Center of Canister as a Function of Time

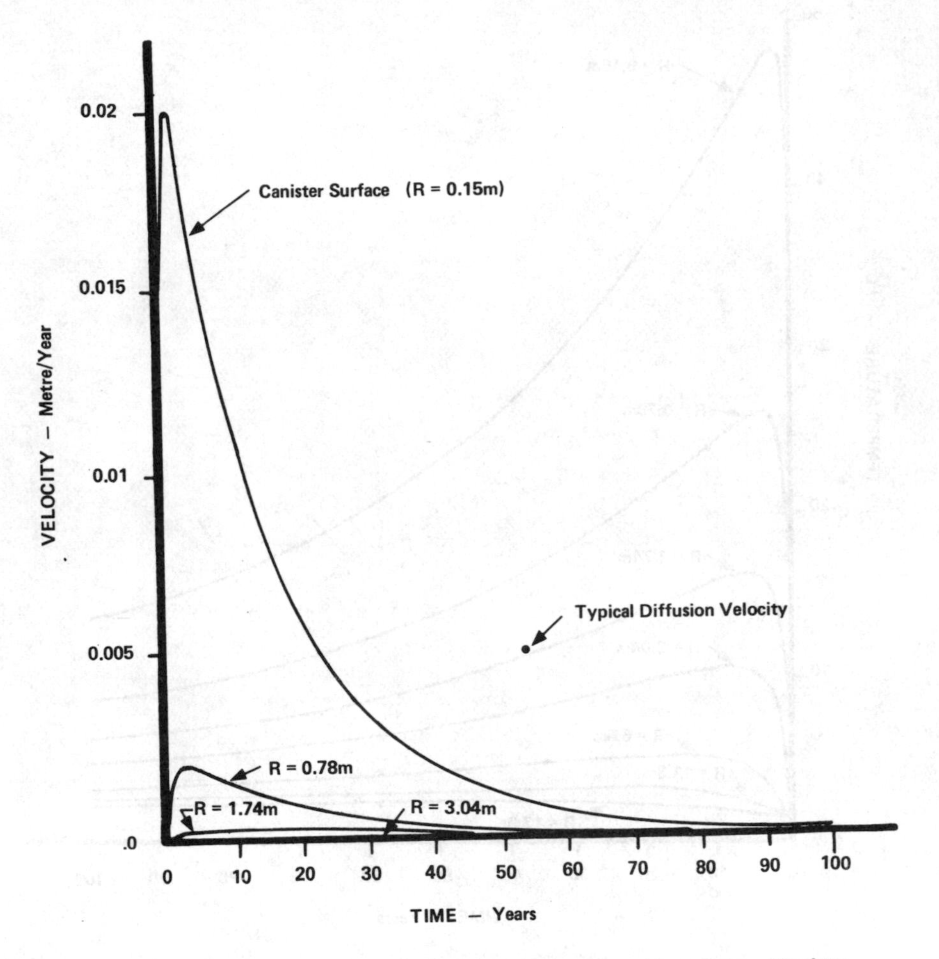

FIGURE 5. Velocity at Three Locations Radially Outward from Center of Canister as a Function of Time, 1.5KW Initial Thermal Power, HLW

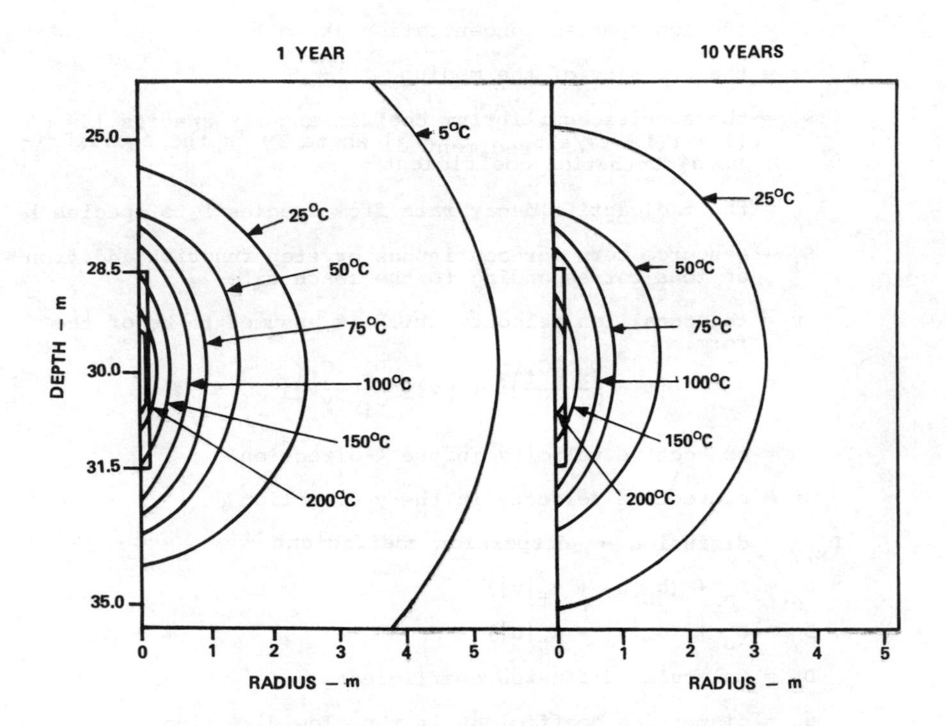

FIGURE 6. Isothern Patterns Near Canister at Two Times for 1.5KW Initial Thermal Power, HLW

effects are not treated in that species are injected outside of the thermally altered zone as volumetric source terms.

The equations describing the migration of contaminant ions in a porous saturated bed are of the form:

$$\frac{\partial (C_i K_i)}{\partial t} + \nabla \cdot (C_i \vec{v}) = -\sum_{k=1}^{N} (\lambda_{ik} K_i \varepsilon C_i) + \sum_{k=1}^{N} \lambda_{ki} K_k \varepsilon C_k + S_i \quad , \tag{7}$$

where,

C_i = the ion species concentration (kg/m^3)

ε = the porosity of the medium

K_i = the species equilibrium coefficient for species i = $[1 + ((1-\varepsilon)/\varepsilon)\rho_{sediment} K_d]$ where K_d is the equilibrium distribution coefficient

λ_{ik} = the radioactive decay rate from species i to species k

S_i = a source term for continuous or step function addition of ions corresponding to the leach rate

$\vec{v}$ = the total ion velocity which is assumed to be of the form:

$$\left(u - \frac{D_x}{C_i}\frac{\partial C_i}{\partial x}\right)\vec{i} + \left(v - \frac{D_y}{C_i}\frac{\partial C_i}{\partial y}\right)\vec{j}$$

u = convective velocity in the x-direction

v = convective velocity in the y-direction

$D_{x,y}$ = diffusion + dispersion coefficient

$D_x = D_o + (\alpha_L |u| + \alpha_t |v|)$

$D_y = D_o + (\alpha_L |v| + \alpha_t |u|)$

D_o = molecular diffusion coefficient

α_L = dispersion coefficient in the flow direction

α_t = dispersion coefficient transverse to the flow

The equilibrium distribution coefficient, K_d, is a function of species concentration (4) and is assumed to be of the form:

$$K_d = \frac{k_2}{1 + k_1 C} + \frac{k_4}{1 + k_3 C} \quad . \tag{8}$$

Equation 7 was formulated in finite difference form for solution.

The computational domain is a region necessarily of the same configuration shown in Figure 1 since temperature and velocity from the MARIAH calculations are used in IONMIG. The boundary conditions used for the IONMIG calculations are underlined in Figure 1. The side and bottom boundaries are impermeable to the nuclides. At the ocean-sediment interface it is assumed that nuclides are removed as

(4) Provisions for K_d and D_i as functions of temperature are currently being incorporated.

they are introduced so $C_i = 0$. The concentrations and sources are initially specified throughout the field and the code computes the subsequent time variations.

The IONMIG code has been used to make initial breakthrough, concentration distribution and surface injection estimates for the study case of Figure 1. For the present, it is assumed that the majority of the radioisotope inventory in nuclear waste will migrate as cations. Pu and Cs were chosen as being typical: ^{239}Pu -- typical of the transuranic elements with long half-life and high K_d; ^{137}Cs -- typical of fission products with a relatively short half-life and moderate K_d. However, concern has been expressed regarding the behavior of anionic species and so ^{129}I -- a long half-life element with low K_d and ^{99}Tc -- a fission product with long half-life and low K_d, were included for reference purposes. ^{129}I is present in SF or liquid waste but usually not in solidified HLW since it volatilizes during processing. Owing to the toxcicity, ^{129}I cannot be released and must be isolated and disposed of in some manner; most probably in the same geologic repository that receives the HLW. Thus, it poses essentially the same containment problem as if included with the HLW. The nuclides considered in the decay chain with each of the four elements are listed in Table II. The initial canister inventories were taken from the ORIGEN [8] calculations made for the thermal source terms.

Additional data used in the calculations are given in Table III. Data given by Li and Gregory [13] were used in estimating the value for D_o. For lack of more complete data, D_o was assumed the same for each species. In the absence of complete data, the dispersivity coefficients (α_L, α_t) were estimated from some "in-house" information. Rough estimates of dispersivity are acceptable since the velocity is very low and decays to zero early in the problem making dispersion a negligible effect. The K_d correlations for the sediment, taken from the ongoing work of Erickson [14], are given in Table III. The data are from batch experiments. In the tests, the Pu was in the +4 valence state and Cs in the +1 state.

In defining the source term for the IONMIG calculations, the canister inventory of each element was assumed to be initially uniformly distributed within a 5 m long by 2.0 m radius cylinder about the canister. The time delay required for the canister to corrode and for the material to leach out of the waste form was neglected. In addition, effects of the relatively high temperature region near the canister are neglected. This is also reasonable since the region is relatively small and the time for thermal decay and for heat dissipation is negligible relative to the migration time.

Calculations for ^{137}Cs were run out to 3500 years. Little motion took place prior to complete decay of the ^{137}Cs into ^{137}Ba. Since the half-life of ^{137}Cs is only 30 years, virtually all the ^{137}Cs decayed in place, and essentially none reached the surface. The same behavior is expected of other short-lived species with moderate values of K_d such as ^{90}Sr.

For long-lived isotopes having a high K_d, such as ^{239}Pu, the situation is similar except the time scales are large. Figure 7 shows the distribution at 100,000 years of ^{239}Pu over a vertical symmetry plane containing the buried canister. Although 4.1 half-lives of ^{239}Pu have elapsed, the remaining plutonium is almost one-third of the original inventory because of the decay of ^{243}Am to ^{239}Np and then to ^{239}Pu. In this calculation, the Am and Np were assumed to have the same K_d as the plutonium. Data currently available indicate this is a good assumption. The migration rate of plutonium is so slow (a few metres per 100,000 years) that even after 10^6 years breakthrough has not occurred. At this time, the concentration had decayed to 10^{-10} gm.

TABLE II

NUCLIDE DECAY CHAINS AND ZERO TIME CANISTER INVENTORY FOR IONMIG CALCULATIONS HIGH LEVEL WASTE 1.5 kW INITIAL THERMAL POWER

Radionuclide	Initial Canister Inventory (kg)
^{239}Pu	3.58×10^{-2}
^{247}Bk	5.3×10^{-7}
^{243}Am	0.121
^{239}Np	1.0×10^{-7}
^{243}Cm	6.92×10^{-5}
^{235}U	1.0×10^{-5}
^{129}I	0.305
^{129}Xe	1.178×10^{-5}
^{99}Tc	1.13
^{99}Ru	4.32×10^{-5}
^{137}Cs	1.33
^{137}Ba	2.01×10^{-7}

TABLE III

PROPERTIES FOR NUCLIDE MIGRATION ANALYSIS - IONMIG

D_o	m^2/year	0.01
α_L	m	6.1
α_t	m	0.61
$K_{d_{Pu}}$	m^3/kg	$\frac{100}{1.0 + 3 \times 10^8 C_{Pu}} + 0.01$
$K_{d_{Cs}}$	m^3/kg	$\frac{10}{1.0 + 2 \times 10^5 C_{Cs}} + 0.1$
$K_{d_{I,Tc}}$	m^3/kg	$\frac{0.0001}{1.0 + 1 \times 10^4 C_{I,Tc}}$

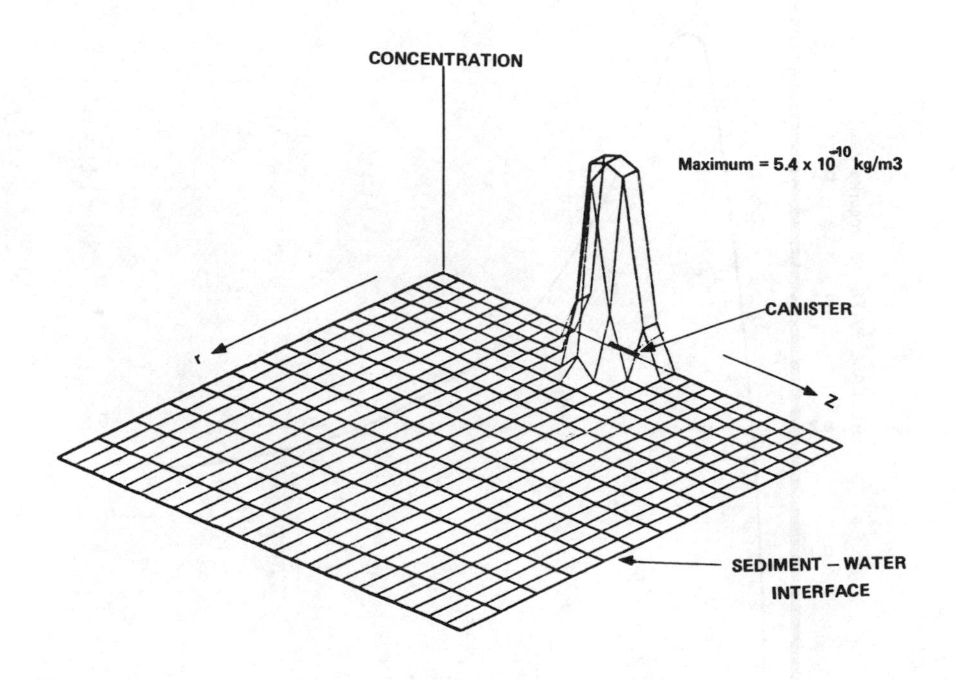

FIGURE 7. Plutonium Concentration in the Sediment at 100,000 Years

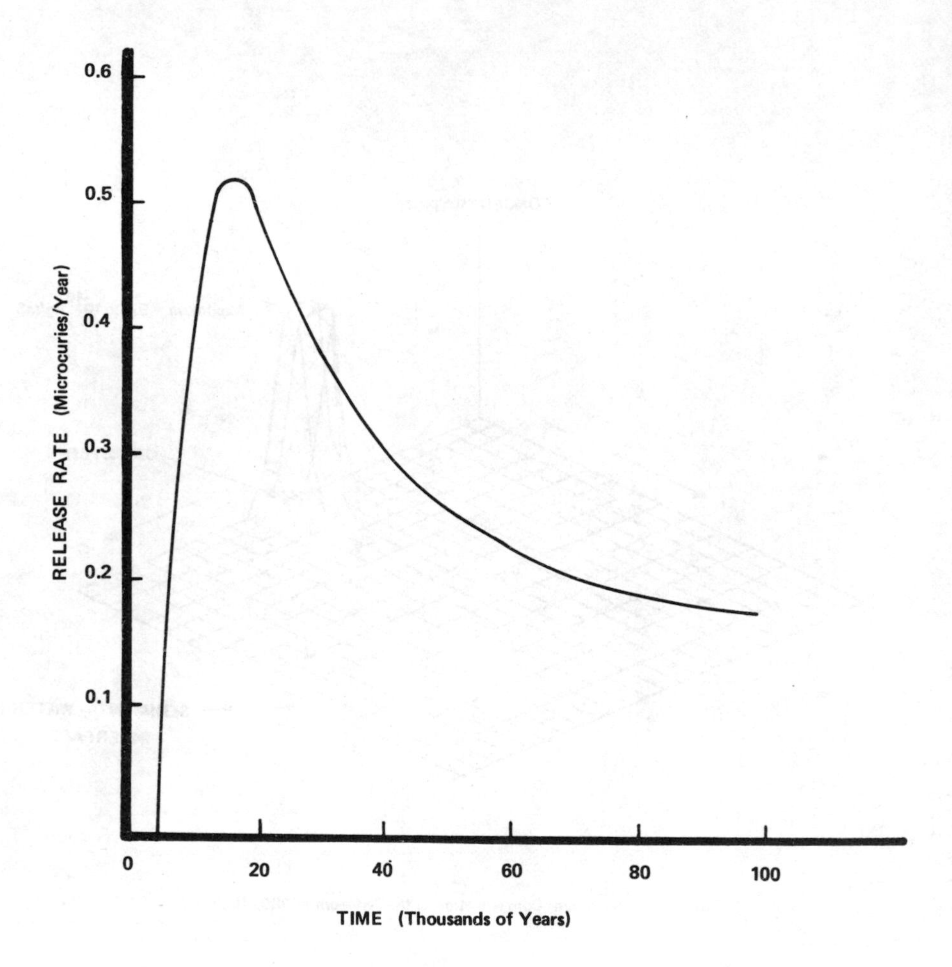

FIGURE 8. Release Rate of ^{129}I from the Sediment Surface Covering a Single Canister at 30m Depth

For nuclides having a long half-life and a very small K_d, such as ^{129}I and ^{99}Tc, the behavior is quite different. Without the retarding action of sorption, these substances break through the sediment in a relatively short time (5000 years). Figure 8 shows the release rate of radioactivity associated with the decay of ^{129}I as a function of time. The release rate reaches a peak of 0.52 μCi/year at about 15,000 years and then declines to approximately one-fourth of that value in 100,000 years. ^{129}I has a low specific activity, decaying to ^{129}Xe by beta decay with a half-life of 15.9 million years. Figure 9 shows a similar release rate profile for ^{99}Tc. The peak value of 180 μCi/year is much higher than that for ^{129}I because both the initial canister inventory and the specific activity of ^{99}Tc which beta decays to ^{99}Ru with a half-life of 213,000 years, are higher.

Based on these calculations, it appears that but a miniscule fraction of those radioisotopes migrating as cations will ever cross the sediment-water interface. This is not the case for Tc and I, elements which will travel in solution as anions. Care will, therefore, be required to reduce their release rate to values that are comparable to the cationic release rates.

The release rates of ^{129}I and ^{99}Tc can be put somewhat into perspective by comparing them to the current IAEA allowables [15]. Under the special conditions outlined in the IAEA convention, the upper limits on activity release rates from all sources (other than natural sources) when released into an ocean basin with a volume of not less than 10^{17} m^3 shall not exceed 10^8 Ci/year for β/γ emitters with half-lives of at least 0.5 year (excluding tritium and emitters of unknown half-life). In the present case, the peak ^{99}Tc release rate is 180 μCi/year. To approach the convention limit would require some 5.5×10^{11} canisters to be emplaced! This number of canisters is many orders of magnitude greater than any present projection. One can reason that the ^{99}Tc injection rate is low by present standards and the ^{129}I is even lower.

CONCLUSIONS

Based on the analysis to date and the available sediment property data, we conclude that for the seabed sediment:

(1) When the canister initial power is limited to 1.5 kW in order to keep the maximum sediment temperature between 200°C and 250°C as presently dictated by physical, thermochemical, and canister corrosion considerations, the high temperature region ($T > 100°C$) surrounding the waste canister is of limited extent in both space and time (0.8 m radius x 3.6 m long; less than 25 years).

(2) Total fluid displacement due to the convective velocity induced by the thermal energy is small (~3%) relative to the proposed canister burial depths.

(3) Migration calculations using presently known sediment properties for representative nuclides ^{137}Cs, ^{239}Pu, ^{129}I, and ^{99}Tc indicate that: (a) fission products such as ^{137}Cs and ^{90}Sr will decay in place with no release, (b) long-life nuclides such as ^{239}Pu diffuse so slowly through the sediment that release is delayed for such extremely long times that the release rates are negligible, and (c) nuclides that migrate as anions with a long half-life and a very small sorption coefficient such as ^{129}I and ^{99}Tc break through the sediment in several thousand years, but the release rates and specific activity are very low.

(4) Nothing has been revealed by thermal, fluid, and mass transport analyses performed to date to indicate that the red clay sediments cannot be an effective containment barrier for nuclear waste.

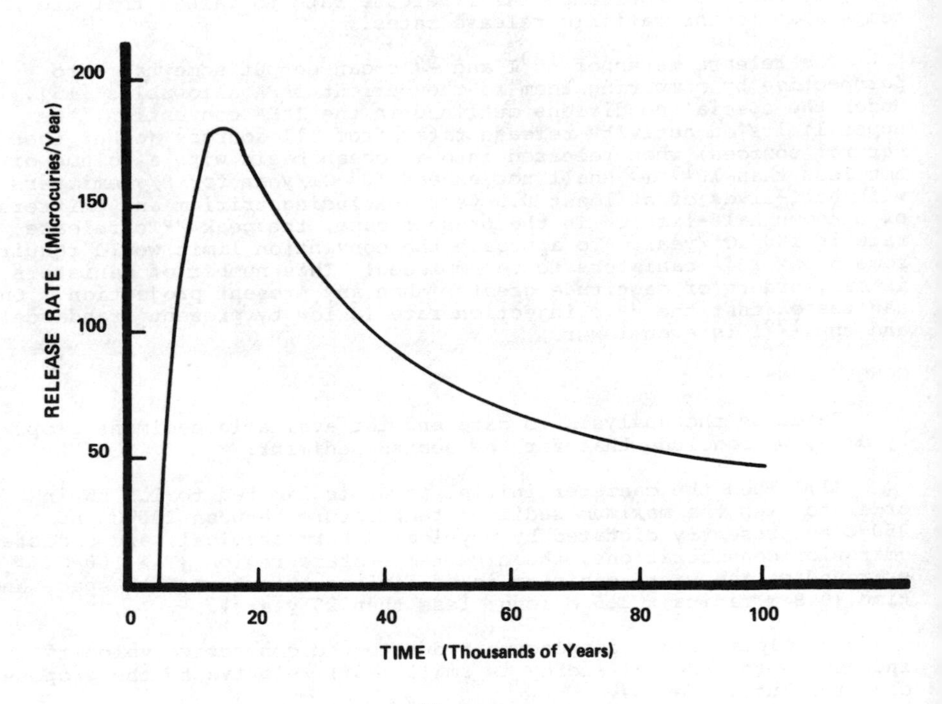

Figure 9 , Release Rate of ^{99}Tc from the Sediment Surface Covering a Single Canister at 30m Depth

REFERENCES

1. "High Level Wastes in the Seabed?" Oceanus, Vol. 20, No. 1, Winter 1977.

2. "Seabed Disposal Program, Annual Report, January to December 1977, Volume I and II," SAND78-1359, Sandia Laboratories, Albuquerque, NM, Jaunary 1979.

3. "Seabed Disposal Program, Annual Report, January to December 1978, Volume I and II," SAND79-1618, Sandia Laboratories, Albuquerque, NM, to be published.

4. J. L. Krumhansl and G. R. Hadley, "Thermal and Chemical Properties of Saturated Marine Clays in a Near Field Environment," Presented at the Workshop on Argillaceous Materials for Isolation of Radioactive Waste, OECD Nuclear Energy, Paris, France, September 9-12, 1979.

5. D. K. Gartling and C. E. Hickox, "MARIAH--A Finite Element Computer Program for Incompressible Porous Flow Problems, Part I--Theoretical Background," SAND79-1622, Sandia Laboratories, Albuquerque, NM, to be published.

6. D. K. Gartling and C. E. Hickox, "MARIAH--A Finite Element Computer Program for Incompressible Porous Flow Problems, Part II--User's Manual," SAND79-1623, Sandia Laboratories, Albuquerque, NM, to be published.

7. J. R. Philip, "Transient Fluid Motion in Saturated Fluid Media," Australian Journal of Physics, Vol. 10, pp. 43-53, 1957.

8. M. J. Bell, "ORIGEN--The ORNL Isotope Generation and Depletion Code," ORNL-4628, Oak Ridge National Laboratory, May 1973.

9. A. J. Silva and D. I. Calnan, "Geotechnical Aspects of Sub-Surface Seabed Disposal of High Level Radioactive Wastes," Annual Report No. 5, Dept. of Ocean Engineering, University of Rhode Island, January 1979.

10. J. W. Braithwaite and M. A. Molecke, "High-Level Waste Canister Corrosion Studies Pertinent to Geologic Isolation," Nuclear Waste Management and Technology, Vol. 1, No. 1, Pergamon Press, 1979.

11. C. E. Hickox and H. A. Watts, "Steady Thermal Convection from a Concentrated Source in a Porous Medium," 79-HT-69, ASME/AIChE, 18th National Heat Transfer Conference, San Diego, CA, August 1979.

12. A. J. Russo, "Prediction of the Migration of Several Radionuclides in Ocean Sediment with the Computer Code IONMIG, A Preliminary Report," SAND79-1666, Sandia Laboratories, Albuquerque, NM, to be published.

13. Yuan-Hui Li and Sandra Gregory, "Diffusion of Ions in Sea Water and Deep-Sea Sediments," Geochimica et Cosmochimics Acta, Vol. 38, pp. 703-714, 1974.

14. Personal communication, K. L. Erickson, Division 5812, Sandia Laboratories, Albuquerque, NM.

15. "Convention on the Prevention of Marine Pollution by Dumping of Wastes and Other Matter," INFCIRC/205/Add 1/Rev 1, International Atomic Energy Agency, August 1978.

Discussion

A.G. DUNCAN, United Kingdom

Referring to your Figure 7, what benefit derives from containing the waste within the canister for, let's say, 1000 years, the period during which thermal convection might occur ?

D.F. McVEY, United States

Thermal convection makes a small contribution to the total nuclide motion. Hence, a canister of 1000 years life would have little effect on the data in Figure 7 ; mainly the time scale would be shifted by 1000 years. However, let me emphasize, I feel that the canister should be designed to have as long a life as possible. The minimum life should be at least three thermal half-lives for high level waste.

J.B. LEWIS, United Kingdom

While I agree that the amounts of radioactivity likely to be released to the ocean basin are small, I think that the comparison with the IAEA limits for low level sea disposal is naive. The IAEA limits refer to an upper dumping limit of 100,000 tonnes/year, a figure not remotely approached. A better comparison would be with total amounts of radionuclides released to the marine environment by fallout.

D.F. McVEY, United States

I agree. In the time we had to write the paper we could not find the required fallout information. We intend to include these comparisons in future publications. Our main point is that the release rates would be very small.

F. GERA, NEA

For the sake of perspective it would also be useful to compare released activity with the activity of existing natural radioactive elements in the ocean.

D.F. McVEY, United States

Yes, as we get the appropriate data we will.

F. GERA, NEA

Have you considered the possibility that the thermal gradient might induce sufficient flow to fluidize the sediment and generate convection cells ? Do you have an idea of the temperatures that might cause such phenomenon ? If it is physically possible ?

D.F. McVEY, United States

Yes ! We are developing an analysis to address the water motion in more detail. This work is presently being done by Dr. Paul Dawson at Cornell University. This analysis allows pore water motion, and in contrast to MARIAH code, also permits motion of the mineral matrix. Data for the constitutive equations for this code are being developed at the University of Rhode Island and at the U.S. Waterways Experiment Station. This code, along with appropriate experiments, should answer the questions you have posed.

D. DIRMIKIS, United Kingdom

Are the computer codes described in your paper coupled together ?

D.F. McVEY, United States

No, they are not at the present time. Given the unknowns in the experimental data (Kd, permeability, effect of moving samples from the ocean floor to the laboratory) the errors in running a non-coupled problem are most likely of second order, particularly between the thermal and ion migration. It will undoubtly be necessary to couple the thermal and structural (or mechanical) response. It may also be desirable to couple the thermal, thermochemistry and ion migration calculations, when the data are of sufficient scope and accuracy to warrant it.

SEDIMENT CHARACTERISTICS OF THE 2800 AND
3800 METER ATLANTIC NUCLEAR WASTE SITES
APPLICABLE TO ISOLATION OF RADIOACTIVE WASTE

James Neiheisel
Environmental Protection Agency
Washington, D.C., United States of America

ABSTRACT

The sediments of the Atlantic nuclear waste sites at 2800 and 3800 meter depth located along the continental rise off the east coast of the United States have been analyzed for sediment texture, mineral composition, physical properties, and chemical parameters of value in predicting the radionuclide retention capabilities of the sediment. The 2800 meter site occurs on a gently sloping, depositional surface and consists of silty-clay with one-third clay minerals (illite, kaolinite, chlorite, and montmorillonite) and moderate cation exchange capacity (16-25 meq/100g). The 3800 meter site occurs in the main axis of Hudson Submarine Canyon with similar parameters except for local slump blocks predominantly montmorillonite clay fraction and cation exchange capacity of 43.8 meq/100g. Radionuclide retention by sediment is greatest in buried waste drums free from exposure or bioturbation in the sediment.

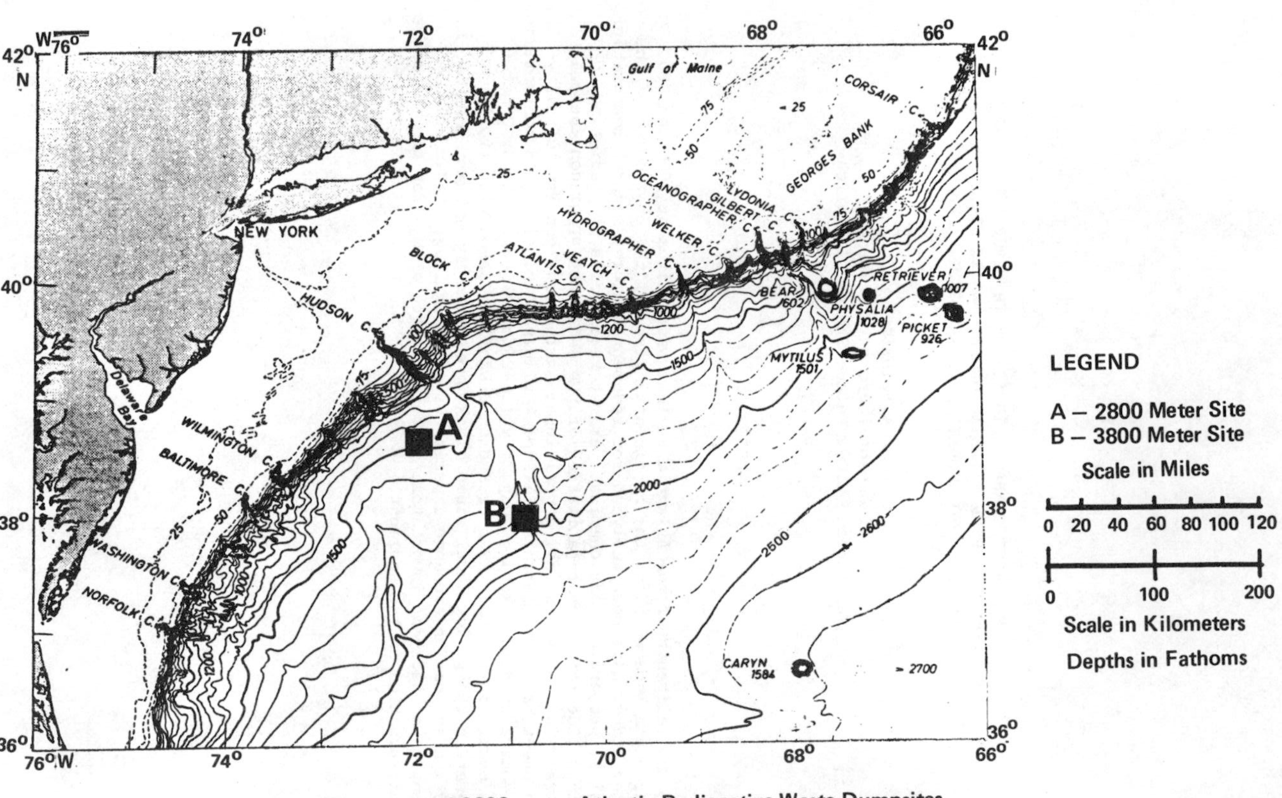

Figure 1. Location of the 2800 meter and 3800 meter Atlantic Radioactive Waste Dumpsites on the Atlantic Continental Rise.

INTRODUCTION

Argillaceous materials have physical and chemical properties which are desirable for the isolation of radioactive waste. These materials occur in the sedimentary rocks on the continents and in the sediments of the oceans. Although other candidate lithologies are being evaluated for isolation of nuclear waste to allow decay to innocuous levels prior to reentry into the biosphere, argillaceous materials must be investigated as an alternative for their potential capabilities in this complex problem.

The United States Environmental Protection Agency, pursuant to Public Law 92-532, the Marine Protection Research and Sanctuaries Act, is engaged in an intensive program of site-characterization studies of abandoned nuclear waste dumpsites in argillaceous sediments of the Atlantic and Pacific Oceans. The major portion of 76,300 steel drums of low level nuclear waste were dumped at two Atlantic sites and two Pacific sites between 1946 and 1962 prior to shallow land burial practices on the continents. Information from these investigations will provide basic data of value in assessing the argillaceous sediment of the ocean environment in their capabilities to act as a "sink" to the radionuclides released from the waste form.

This paper presents site specific information on the 2800 meter and 3800 meter Atlantic nuclear waste sites containing approximately 40 percent of the nuclear waste deposited in the oceans by the United States. Consideration is given to the geologic setting, sediment texture, physical properties, mineral composition, and chemical parameters of the waste sites. The sediment is viewed in its potential role to act as a barrier to radionuclide migration and mechanisms delineated that retard such capabilities.

GEOLOGIC SETTING

The Atlantic nuclear waste dumpsites comprise areas of approximately 256 square kilometers with centers at 38° 30' N, 72° 06' W and 37° 50' N, 70° 35' W (Figure 1). The 2800 meter site occurs on the upper continental rise approximately 190 kilometers off the New Jersey coast. The 3800 meter site is situated on the lower continental rise approximately 310 kilometers off the New York coast.

The 2800 meter site bottom topography has been described by Rawson and Ryan (1) as gently sloping, uniform terrain , carpeted with fine grained sediments. Steel 55 gallon nuclear waste drums only partially buried and containing plumes of sediment on their lee sides reflect weak prevailing bottom current from the northeast parallel to bottom topography. Rawson and Ryan (1) estimate sedimentation rate in this area, to be 6.8 cm per 1000 years for the recent Epoch (last 11,000 years).

The 3800 meter site is situated in the main axis of the Hudson Submarine Canyon near the confluence with Block Submarine Canyon (Figure 1). Hanselman and Ryan (2), from observation in the Alvin, describe the walls of the canyon rising from the 1 kilometer wide canyon floor as gently sloping to vertical and about 200 meters high. The canyon floor contains claystone slump blocks of various sizes and shapes in a fine grained sediment. Channel thalwegs in contact with channel walls apparently provide the cutting mechanism that activates slumps or gravity slides (2). The bottom currents operating in this waste site are stronger than those at the 2800 meter site. Nuclear waste drums deposited in this environment could be moved, buried, or burst by slides or slumps from the channel walls.

SEDIMENT SAMPLING

The sediment sample locations at the 2800 meter and 3800 meter Atlantic sites are depicted in Figure 2. The samples at the 2800 meter site consist of grab samples from surface ships (SH20, SH25, and SH26), and core samples from Alvin dives with cores taken to 30 cm depth in proximity to nuclear waste drums. The latitude and longitude of the three Alvin dives taken

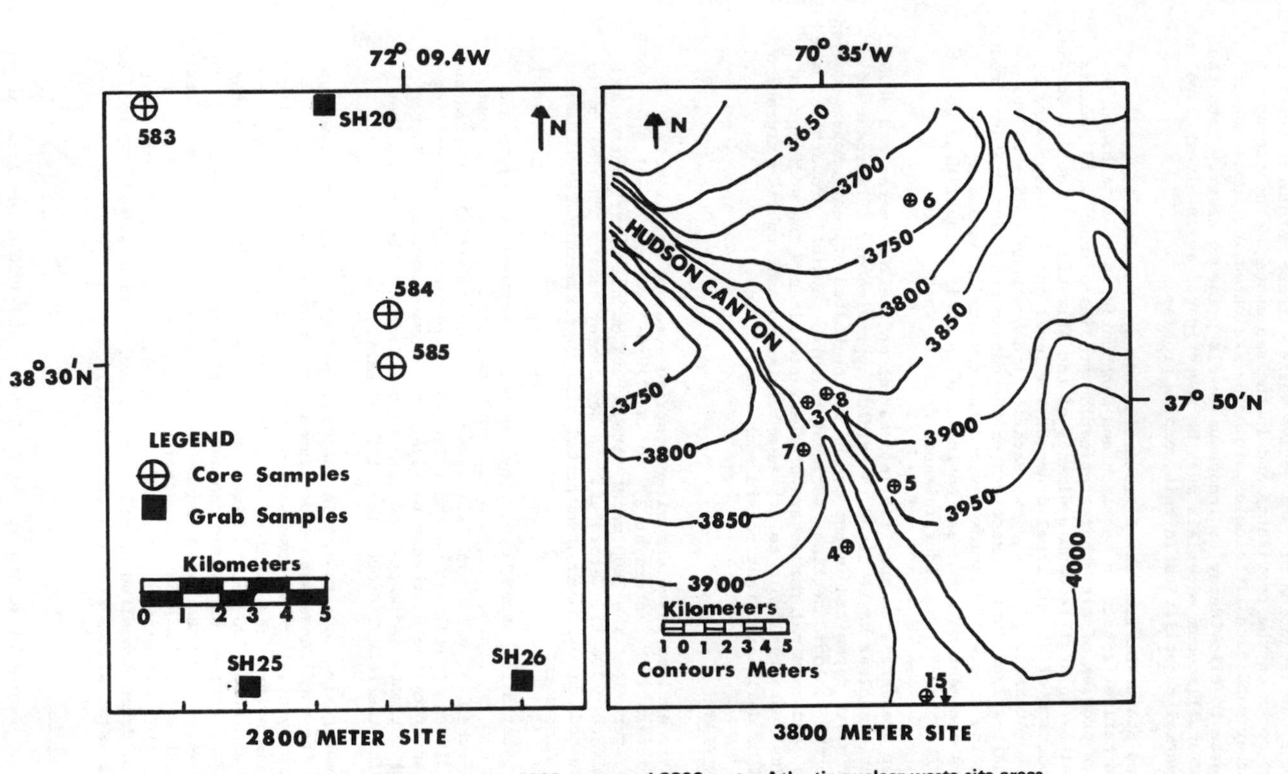

Figure 2. Sample locations within the 2800 meter and 3800 meter Atlantic nuclear waste site areas.

during July 1975 are as follows:

Dive 583	38° 33.6' N	72° 12.4' W
Dive 584	38° 30.7' N	72° 09.4' W
Dive 585	38° 30.2' N	72° 09.4' W

The sediment samples from the 3800 meter site were taken with Soutar Box Corer from the SS ADVANCE II during June 1978. The geographic coordinates of these sample locations, depicted in Figure 2, are listed below:

Field Location 2	37° 50.83' N	70° 35.50' W
Field Location 3	37° 49.30' N	70° 36.71' W
Feild Location 4	37° 45.03' N	70° 35.75' W
Field Location 5	37° 46.75' N	70° 34.01' W
Field Location 6	37° 54.65' N	70° 32.69' W
Field Location 7	37° 48.10' N	70° 37.11' W
Field Location 8	37° 49.79' N	70° 36.13' W
Field Location 15	37° 43.81' N	70° 32.38' W

Sediment tube cores used for determination of sediment properties were obtained as subcorings from the Soutar Box Corer. Samples number 7 and number 5 appeared desiccated and contained numerous bore holes from organisms, whereas, the other more typical samples were more plastic and less consolidated.

METHODS OF ANALYSIS

The sediments were analyzed pursuant to an Interagency Agreement with the U.S. Army Corps of Engineers. Mineral identification was performed by the writer, cation exchange capacity by Dr. Kevin Beck, and supporting scanning electron microscopy and physical testing by the USCE, South Atlantic Division Laboratory and Georgia Tech. Engineering Experiment Station.

The grain size analysis, bulk specific gravity, porosity, and Atterberg Limits tests were performed in accordance with standard procedures (EM-110-2-1906 (3)); x-ray diffraction analysis of the silt and clay size fractions were performed in accordance with the methods of Biscaye, 1965, (4). The special techniques necessary to distinguish kaolinite from chlorite on x-ray diffractograms of the clay fraction were in accordance with Biscaye, 1964 (5), utilizing the slow scan and measuring areas of the 3.54A chlorite and 3.58A kaolinite peaks. The mineral identification of the sand size fraction was determined by standard petrographic techniques. Carbonate evaluations were made of each sieve size by acid leach techniques using 1:4 HCl. The fractional components of each size were weighed by the grain size curve and the sum recorded as the average in the sample.

The cation exchange capacity determination were made using a method similar to that of Zaytseva, as briefly described by Sayles and Mangelsdorf (6). The samples were squeezed in a stainless steel device, the pore water was collected and analyzed for Na+, K+, Mg++, and Ca++, and the remaining squeeze cake was split into two parts. One part was used for determinations of the residual water content (110°C drying) and the other was leached of residual sea water and exchange cations, using a succession of washes (80% methanol, 1 N NH_4Cl, adjusted to pH 8 with NH_4OH). The exchange cations were calculated by subtracting the seawater contribution from the total leach solution. The purpose of this involved procedure was to circumvent the exchange cation-seawater reequilibrium (Donnan effect) that occurs during the washing step which proceeds the exchange in the more traditional approach. The pH adjustment and the use of methanol were used to minimize $CaCO_3$ solution during leaching.

SEDIMENT TEXTURE

The relative percentages of the sand, silt, and clay fraction of the sediment from the 2800 and 3800 meter site is depicted in the ternary plots of Figure 3. The descriptive nomenclature for the sediment is in accordance with Shepard's classification (7). The majority of the sediment

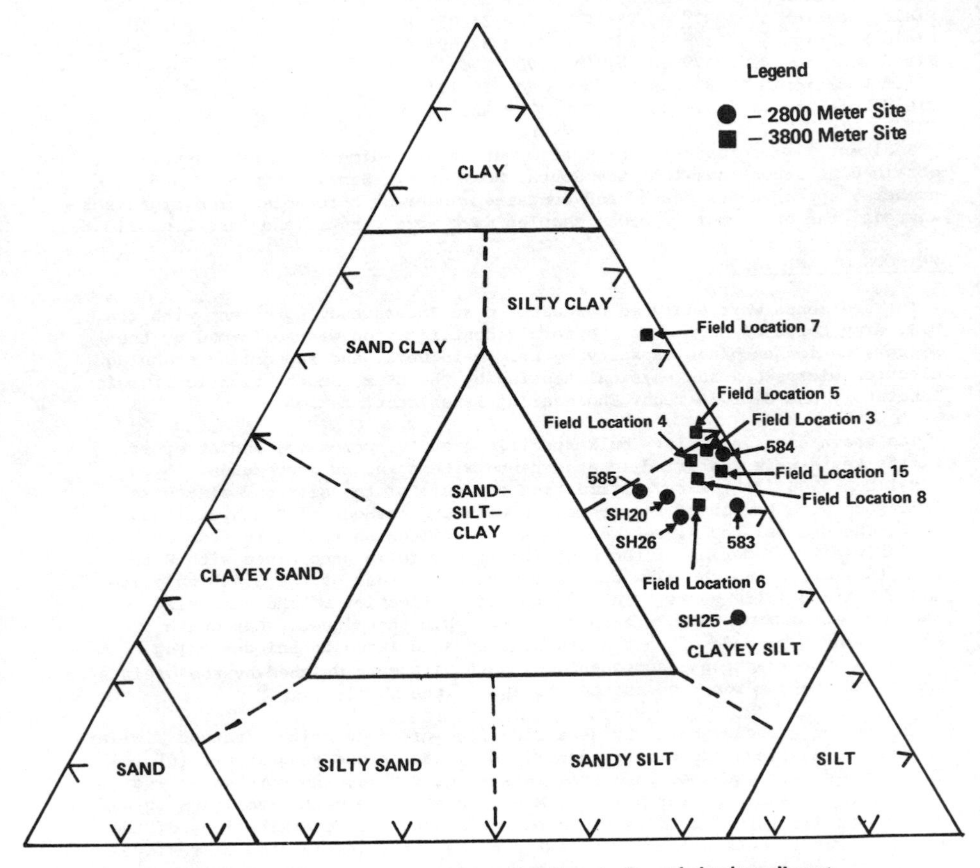

Figure 3. Ternary plot depicting relative percentages of sand, silt, and clay in sediment samples from the Atlantic nuclear waste sites. Nomenclature is in accordance with Shepard.

at both sites is a clayey silt. Finer textured silty clay occurs at the 3800 meter site in the samples from field locations 5 and 7 (Figure 3).

The fine grained texture of the sediment as well as the relatively high clay mineral content will be shown to be favorable factors as regards radionuclide retention by the sediment.

SEDIMENTARY PARAMETERS

The median diameter, standard deviation, skewness, and kurtosis are parameters commonly used in most sediment investigations. These parameters are calculated from the cumulative grain size distribution curves. Although these parameters have more meaningful application in the higher energy zones of shallower depth, they will be briefly considered in this investigation for the limited use in correlating the sediment properties.

The median diameter is the 50 percentile of the grain size distribution curve. The median diameter for the samples from the 2800 meter Atlantic waste site is markedly uniform over the entire area with a range between 0.002 and 0.005 mm and an average of 0.003 mm. The samples at the 3800 meter site range between less than 0.001 mm for sample number 7 to slightly in excess of 0.003 mm with the general average (exclusive of field sample 7) approximately 0.003 mm. Field sample number 7 appears atypical of the other samples in all the parameters tested and probably represents a slump block from the canyon wall.

The standard deviation is a measure of uniformity or statistical measurement of sorting in the sediment. Arbitrary limits are used by various investigators to describe the standard deviation. In accordance with Friedman's classification (8), the range in standard deviation from 2.39 to 4.57 at the 2800 meter site is categorized as very poorly sorted and extremely poorly sorted sediment. The samples at the 3800 meter fall into a similar grouping. Such a classification is typical of silts and clays and the principal use of this parameter appears in reflecting the uniformity of the sediment.

Skewness and kurtosis are useful parameters in reflecting upon uniformity of the depositional environment of the sediment. The skewness is a measure of the direction and degree of overall deviation from symmetry; at the waste sites the sediment possesses a coarse tail portion relative to the finer sizes. The kurtosis is a measure of the peakedness of the curve, i.e., the relationship of the sorting within the main body of the curve to that of the tails. At the waste sites, the sediments are extremely leptokurtic, reflecting that the central portion of the grain curve is better sorted than the tails.

While sedimentary parameters tell little of the adjustment of the sediments to their present environment or of province, these parameters are nevertheless correlative to homogeneous conditions and other site orientated relationships. These parameters enable a relatively inexpensive means of establishing if uniformity of physical conditions prevails within the area investigated.

PHYSICAL PROPERTIES

The bulk specific gravity and porosity are physical properties used in the prediction of the behavior of radionuclides released to the sediments. Several samples were analyzed for these properties from the 2800 and 3800 meter waste sites.

The bulk specific gravity (in place dry density) generally increases with depth. The lowest measured value of 0.36 g/cm^3 occurs in the surface sample of dive 585 at the 2800 meter site; a sample from 12.5 cm depth from the same core had a bulk density of 1.14 g/cm^3. The average bulk specific gravity at the 2800 meter site to 20 cm depth approximates 0.75 g/cm^3. At the 3800 meter site, the contemporary sediment samples approximate 0.83 g/cm^3.

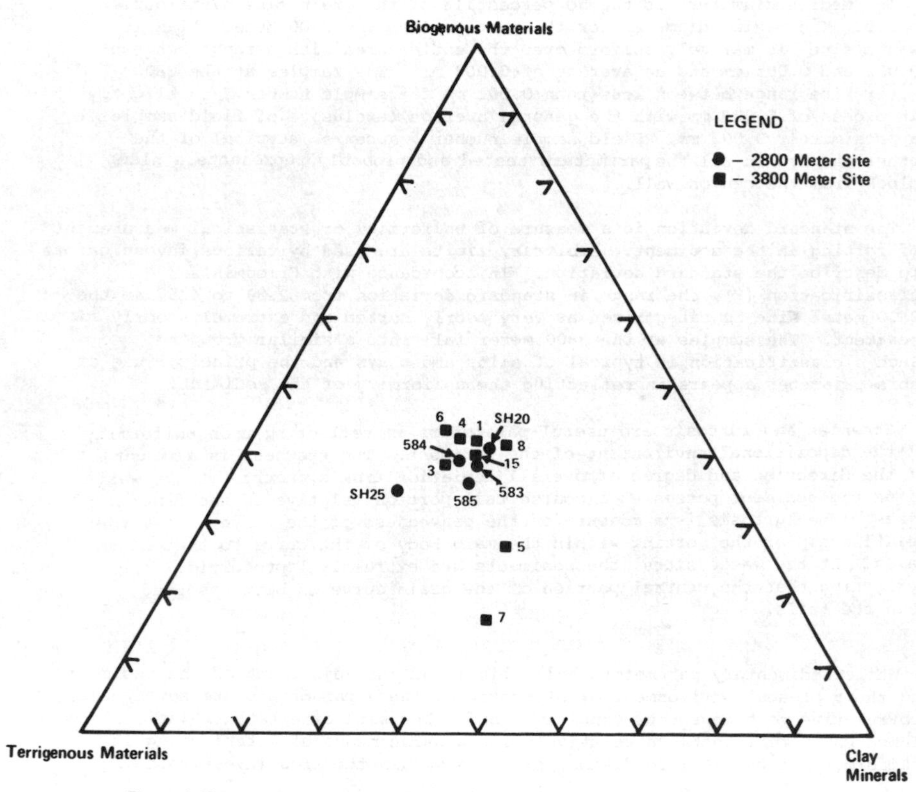

Figure 4. Ternary plot showing relative percentages of biogenous materials, clay minerals, and terrigenous materials in sediment from the 2800 and 3800 Atlantic nuclear waste dumpsites.

TABLE I. Mineralogical and Chemical Parameters Significant to Radionuclide Retention in Sediments of the 2800 and 3800 Meter Atlantic Sites

	Bulk Composition			Clay Mineral Fraction				Cation Exchange Capacity meq/100g				
Sample Number	Biogenous Materials	Terrigenous Materials	Clay Minerals	Illite	Kaolinite	Chlorite	Montmor-illonite	Na^+	K^+	Mg^{++}	Ca^{++}	TOTAL
					2800 Meter Site							
SH 20	42	28	30	52	20	18	10	8.0	1.6	4.3	3.5	17.4
SH 25	36	42	22	51	20	18	9	-	-	-	-	-
SH 26	33	39	28	52	20	20	8	7.0	0.8	4.1	3.3	15.2
583	37	33	30	54	18	18	10	9.4	2.1	5.3	4.0	20.8
584	40	33	27	55	22	13	10	11.3	2.6	6.3	5.2	5.4
585	36	34	30	58	20	14	8	6.9	1.4	3.4	3.3	15.0
					3800 Meter Site							
3	40	31	29	53	19	10	18	15.3	3.9	8.0	6.3	33.5
4	42	29	29	51	24	10	15	13.8	3.6	7.7	6.7	31.8
5	27	33	40	30	10	10	50	-	-	-	-	-
6	43	32	25	47	13	7	33	12.3	3.2	6.3	6.2	30.0
7	16	32	52	5	9	3	83	19.7	6.1	10.1	7.9	43.8
8	43	24	23	53	10	10	27	13.8	3.6	8.3	6.4	32.1
15	40	31	29	53	18	8	21	14.8	4.1	8.3	7.4	35.1

Note: The dive core samples at the 2800 meter site and field core samples at the 3800 meter site are the average of the core samples taken to 20 cm depth.

The porosity, expressed as a fraction of the total sample volume ranges between 0.75 and 0.86 at the 2800 meter site and 0.69 to 0.76 at the 3800 meter site. The slightly lesser average porosity at the greater depth appears to reflect the somewhat higher biogenous carbonate fraction at the 3800 meter site. Both the bulk specific gravity and porosity will be used in approximating radionuclide retention capabilities of the sediment for buried nuclear waste at these waste dumpsites.

Atterberg Limit tests conducted on several samples from both waste sites are relatively inexpensive tests for correlation of physical properties of the sediment. These tests for liquid limit (LL), plastic limit (PL) and Plasticity index (PI), in themselves, mean little but they are useful as indices to significant properties of the sediment. The liquid limit, denoting the water content at which the sediment closes with standard mechanical manipulations, ranges between 91 and 107 for samples from the 2800 meter site and between 82 and 86 at the 3800 meter site (exclusive of field samples 5 and 7 which are respectively 94 and 100). The plastic limit, which is the water content at which drawn out sediment begins to crumble and break, has a narrow range of 31 to 32 at the 2800 meter site and a broader range between 34 and 47 at the 3800 meter site. The plasticity index, which is the difference between the plastic and liquid limits, represents the range in water contents through which the sediment is in a plastic state and is inversely proportional to the ease with which water passes through the soil. The generally high PI (61-76) at the 2800 meter site as compared to the PI at the 3800 meter site (38-52) reflects a more permeable sediment at the 3800 meter site. Samples 5 and 7 at the 3800 meter site have a PI respectively of 54 and 59 and are probably less permeable than the more typical 3800 meter site samples. These differences between sites appear most related to differences in the mineral composition and the biogenous composition of the sediment.

SEDIMENT COMPOSITION

INTRODUCTION

The sediment composition is important in any assessment of the radionuclide retention capabilities of the sediment. In order to group materials of similar chemical behavior as regards their barrier potential to migration of radionuclides from nuclear waste leachate, the sediment is classified into a biogenous group, terrigenous sand and silt group, and a clay mineral group.

The identification of sand-size material is easily facilitated by microscopic analysis, whereas, the identification of the silt and clay fractions requires a combination of x-ray diffraction, scanning electron microscopy, and chemical techniques. The biogenous, terrigenous, and clay materials identified in the sand, silt, and clay-size fractions is given a weighted average based on the grain size distribution curve. The sum of the weighted averages of the materials on each size fraction is that which is reported for the total sample. The average of the biogenous, terrigenous, and clay materials for the samples at each specific location is listed in Table 1 and depicted graphically in Figure 4.

BIOGENOUS MATERIALS

Biogenous materials, originating from the ocean environment, comprise more than a third of the sediment. These materials consist most abundantly of calcareous foraminifera and coccolith tests with minor occurrences of siliceous diatoms and radiolarians. The calcareous planktonic foraminifera tests of the genus Globigerina comprises the largest constituent in the sand-size sediment (Figure 5), while calcareous planktonic coccolith tests of the algal family Coccolithophoridae account for most of the fine-grained (less than 30 microns in diameter) biogenous material of the lower silt and clay-size fraction (Figure 6).

Figure 5 Photomicrograph (26X) of sand-size fraction of sediment core 7, 5-12 cm depth, from Dive 585 in the 2800 meter Atlantic nuclear waste disposal site. Foraminifera comprises the predominant portion of the sand.

Figure 6. Scanning Electron Micrograph (15,000X) of clay-size Coccoliths from dive location 585 (0-5 cm depth) 2800 Meter Atlantic Nuclear Waste Dumpsite;

The general increase of the carbonate content in seaward direction, due to bigenous material, corroborates the observations of Turekian (9) for the North Atlantic shelf and deeper waters (Table 1). The lesser carbonate in field samples 5 and 7 at the 3800 meter site probably reflects the source of this material as slump from the canyon wall bedrock.

Siliceous biogenous material, based on microscopic and SEM obervations, comprises generally less than one percent of the sediment and occurs in the sand and upper silt sizes. Diatoms are observed at the 2800 meter site and both diatoms and radiolarians at the 3800 meter site.

Tubular arenaceous foraminiferal tests occur in the sediment of field locations 5 and 7 of the 3800 meter site. Pyrite filling is common in these tests from field location 7. This striking difference in faunal biogenic materials , the differing clay mineralogy, and physical properties will be shown to relate to older sediment from the Miocene claystone of the canyon walls.

Laboratory investigations of calcareous materials of the soils and geologic formations of the continents reported on by Ames and Rai (10), suggest that radiostrontium will exchange for calcium under favorable chemical conditions. Presently little is known concerning strontium fixation by ocean sediment and laboratory studies to date have not been in tune with the environment of the ocean.

TERRIGENOUS MATERIALS

Terrigenous materials, as used in this grouping of sediment, include all the nonbiogenous materials except the clay minerals. The clay minerals , while probably largely terrigenous detrital materials, are set into a special grouping because of their special radionuclide retention capabilities to act as a barrier to the migration of radioactive waste. The terrigenous materials are predominatly quartz and feldspar but also include minor amounts of mica (biotite, muscovite, and chlorite) and very minor amounts of detrital heavy minerals and glauconite; the latter comprise generally less than one percent of the sediment and while marine in origin are largely derived from erosion of Cretaceous formations on the adjacent continent.

Special detail to the varieties of feldspar and detrital heavy minerals lend support to the source of sediments supplied to the radioactive waste dumpsites. Mineral provinces have been delineated on the adjacent continental shelf and boundaries between these provinces are essentially perpendicular to the shelf break, as would be expected of distribution of sand by turbidity currents that flowed directly downslope. As indicated by Emery and Uchupi (11) there is no "smearing" of the heavy mineral provinces as would be expected of bottom currents flowing parallel with the contours. Source considerations seem especially warrented for the depositional 2800 meter site and this will be considered in a later section of this report.

The terrigenous materials comprise between 28 and 42 percent of the sediment at the 2800 meter site and 24 to 31 percent of the 3800 meter sediment samples (Table 1 and Figure 4). These materials do not significantly retain radionuclides, however, their ability to do so improves with decreasing grain size.

CLAY MINERALS

The clay minerals comprise approximately two-thirds of the clay-size sediment (material finer than 2 micron size in diameter) at the 2800 and 3800 meter Atlantic nuclear waste sites. Their unique dimensions and high cation exchange capacity present a "sink" potential for the retention of radionuclides . For this reason, the clay mineral group are most important in nuclear waste disposal considerations.

The clay mineral suite for the waste sites are listed in Table 1 and

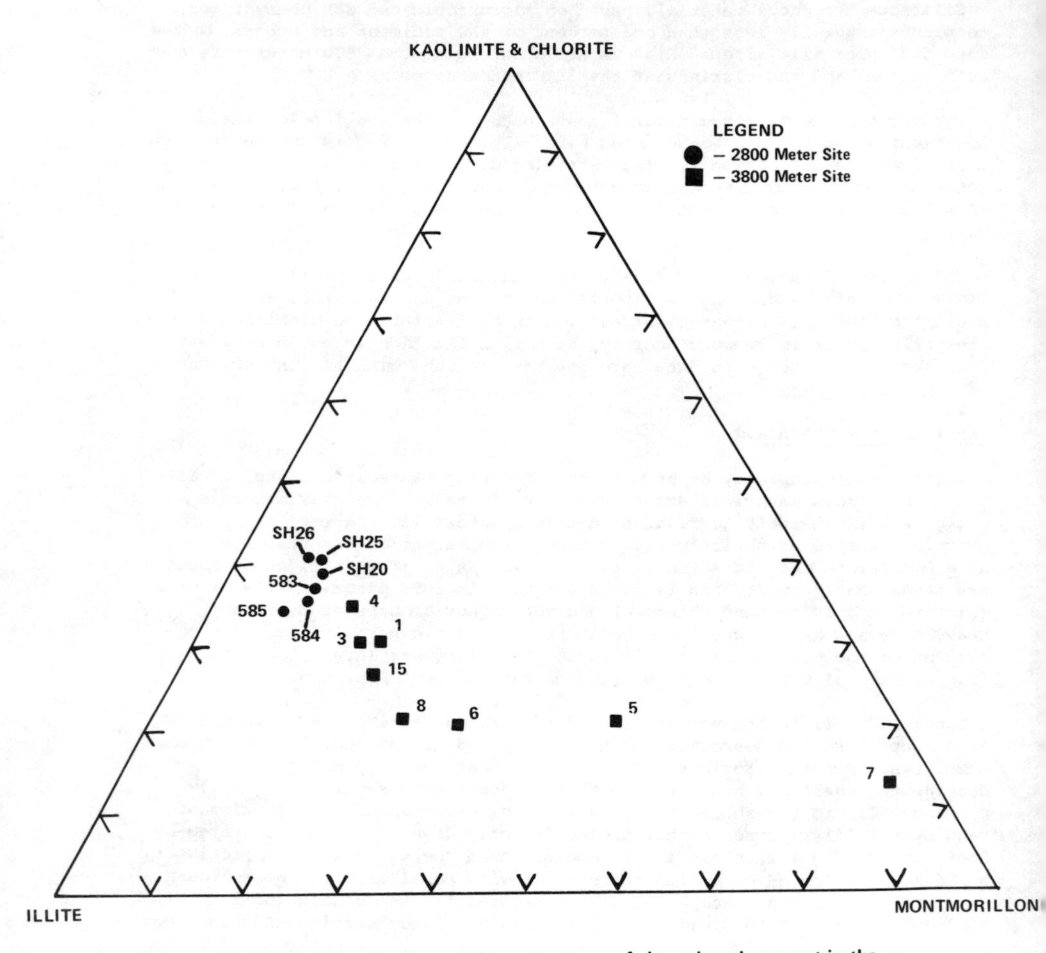

Figure 7. Ternary plot showing relative percentages of clay minerals present in the sediment samples from the 2800 and 3800 meter Atlantic nuclear waste dumpsites.

depicted graphically in the ternary plot of Figure 7. The principal clay minerals in order of abundance are illite, kaolinite, chlorite, and montmorillonite (includes mixed layer montmorillonite - illite or - chlorite) for all the station locations at the 2800 meter site and most of the stations at the 3800 meter site. Field locations 5 and 7 at the 3800 meter site however differ markedly from other samples in being predominantly montmorillonite-rich. These samples are believed to be comprised of clay stone slump material of Miocene age resulting from turbidity flow currents undercutting the banks of the canyon wall. Field samples 6 and 8 at the 3800 meter site probably contain some montmorillonite released to the turbidity flow by boring organisms or scouring processes (Figure 7).

The distribution of the clay minerals in the contemporary sediments of the 2800 meter and 3800 meter sites is in general agreement with the investigations of Hathaway (12) reported on for the sediments of the continental margins of the eastern United States. Although montmorillonite is not important in contemporary sediments in this region, it is abundant in Miocene formations of the continent and continental shelf. The presence of this mineral in any appreciable amounts in sediment along the continental margins of the east coast of the United States is most apt to be related to local eroding areas by turbidity currents or slumping and slide outs such as occur at the 3800 meter site.

CHARACTERISTICS OF CLAY MINERALS

In order to make use of clay minerals in investigations concerning nuclear waste disposal, it is necessary to have a general concept of the chemical and structural make up of the common clay minerals. The following treatment is presented for those readers having little familiarity with the common clay minerals at the Atlantic sites.

The common clay minerals are hydrated silicates comprised of thin sheets held together by predominantly ionic bond; each sheet consists of planes of cations (silica, aluminum, magnesium, or iron) in which the individual cation is surrounded by either four (top sheet) or six (lower sheet) oxygen and hydroxyl ions. The main subdivisions of the clays are based on how these sheets are stacked as follows:

- a. Kaolinite or 1:1 clays contain one silicia sheet (silicon-oxygen tetrahedron) and one sheet of either aluminum, magnesium, or iron (aluminum-oxygen-hydroxol octahedron).
- b. Illites and montmorillonites or 2:1 clays contain two silica sheets which are on either side of an aluminum, magnesium, or iron sheet. In addition, the montmorillonites contain one or two water sheets.
- c. Chlorite or 2:2 clays contain two silica sheets which alternate with two magnesium or iron sheets; the latter have bonds of unequal strength.

Typical clay structures have a thickness of 7 Angstroms as in the case of kaolinite, 10 Angstroms for illite, 14 Angstroms for chlorite and 12 to 14 Angstroms for montmorillonite. Water between montmorillonite tetrahedral layers increases the thickness 2.5 to 4 Angstroms depending on presence of one of two water layers fixed respectively by Ca^{++} and Mg^{++} or Na^{+}. The K^{+} fixed to illite does not add much thickness to the clay structure because it fits into the hexagonal hole in the silica tetrahedron sheets. Chlorite contains no interlayer water but is similar to montmorillonite in size due to a brucite layer. Isomorphous substitution may take place in the clay mineral structure with the amount depending on charge, inoic radius, coordination number or solubility of the participating ions. Common substitution in clay minerals might involve an Fe^{+++} and Al^{+++} substitute for Si^{++++} in the tetrahedral layer while Mg^{++} and Fe^{++} substitute for Al^{+++} in the octahedral layer. The result of a lesser valence cation substitution for a higher valence cation is a negative (-) charge. Most of the charge on kaolinite is usually along the edge, but for montmorillonite is along the surface. Cations, such as Ca++, Na+, H+, Mg++, and K+, are adsorbed on these positions to neutralize the charge when the clay particle encounters a cation-rich environment or receives radionuclides from waste drums. Radioactive waste contains considerable cesium which could substitute for calcium or sodium under proper environmental conditions.

CATION EXCHANGE CAPACITY

The measurement of cation exchange capacity is correlative with the ability of a sediment to adsorb radionuclides from a radioactive waste source. This measurement is expressed in terms of milliequivalents per 100 grams and varies for the different clay minerals. The general range in cation exchange capacity of the common clay minerals, in milliequivalents, is 3-50 for kaolinite, 10-40 for chlorite, 10-40 for illite, and 80-150 for montmorillonite. Thus, sediment rich in montmorillonite would have a higher cation exchange capacity than a kaolinite-rich sediment.

The values for the cation exchange capacity at the 2800 and 3800 meter sites are listed in Table 1. The range is 15.0 to 25.4 meq/100g at the 2800 meter site and 30.0-43.8 meq/100g at the 3800 meter site. As might be expected, the highest cation exchange capacity occurs in the montmorillonite-rich sample from field location 7 which is believed to be slumped claystone from the Miocene age canyon wall.

The presence of organic matter in the sediment could also influence the cation exchange capacity of a sediment. Experiments in recent soils reveal values of 150-500 meq/100g for organic matter. The organic matter at the Atlantic waste sites, however, is probably negligible since values of total organic carbon reported for this section of the ocean floor are on the order of 0.05 percent of the sediment (Emery and Uchupi (11)).

DISTRIBUTION COEFFICIENT, Kd, CONSIDERATIONS

The complex physicochemical reactions that occurs between radionuclides in solution and the ocean sediment is termed sorption. Sorption includes such phenomena as adsorption, ion exchange, colloid filtration, reversible precipitation, and irreversible mineralization. Sorption is expressed in terms of the distribution coefficient (Kd) which is the ratio of the sorbed and dissolved fraction of the radioisotope in the sediment. Knowledge of the Kd of a given radionuclide in sediment, together with information concerning the bulk density and porosity of the *in situ* state, may be used to estimate the retardation factor, Rd, for ground water transport of that radionuclide using the equation $Rd = 1 + Kd\ ^{p}/E$, where p is the bulk density of the medium and E is the porosity. The bulk density used in the equation must have unity of g/cm^3 to obtain a dimensionless Rd. It is interesting to note that the average bulk density cited (0.83 g/cm^3) and average porosity (0.80) nearly cancel each other in the equation for the sediment at the Atlantic nuclear waste sites.

There is not a great deal of relevant sorption data. Kd measurements in the laboratory have been conducted in an oxidizing environment, whereas in reality, ocean sediment is largely a reducing environment. Bondietti and Francis (13) have demonstrated the role of the oxidation state in controlling the solubility of technetium and neptunium from a nuclear waste source; in a reducing environment both radionuclides have high Kd values yet in oxidizing conditions both elements are highly mobile and not readily retained by the geologic media. The pH and Eh are also important factors governing solubility. Both pH and Eh (oxidation-reduction) measurements should be made as soon as the sediment sample is received on board the vessel. Measurements for Kd's in the laboratory should match these measured field parameters.

The pH measured in the laboratory ranged from 7.9 to 8.5 for the 2800 meter site samples and 6.7 to 7.6 for the 3800 meter site samples. The 3800 meter site samples contain some pyrite and any oxidation could have produced a more acid condition than exists at the site.

Heath (14), reporting on preliminary distribution coefficient experiments on montmorillonite-rich Pacific Ocean deep sea sediment, indicates that Kd values range from 3000 to 20,000 for cesium 137 and 100 to 6000 for strontium 90. As contrasted to continental deposits, the marine clays appear to be superior in retaining radionuclides.

In addition to the need for laboratory controlled environmental conditions to match the insitu conditions at a disposal site for accurate Kd assessment, Seitz, et al, (15), in recent column infiltration studies of cesium and other radionuclides on shales and other rock types, has warned of other factors that also influence radionuclide retention. Some of these factors which may have application to ocean sediment considerations include (a) flow rates and dispersive characteristics, (b) nuclide-bearing colloids that react slowly with lithic material, (c) migrating clay particles with adhered radionuclides, (d) chelating of soluble organic compounds, and (e) effects of coexisting species of radionuclides.

SHORT CIRCUITING EFFECTS

If the nuclear waste container is ruptured when deposited or corrodes before it is covered, the radioactive leachate will be released to the waste column for dispersal. However, even if drum burial is achieved before the radioactive leachate is released, there are mechanisms which can short circuit and prevent the sediment to act as a sink for the radionuclides. Bottom dwelling in-faunal organisms can by their movement through the sediment effect bioturbations which act as channels for the migration of radionuclides. Organic matter and organic complexers like EDTA, etc could decrease the retention of some radionuclides and permit their migration through the sediment. The metallic container in oxidizing consumes oxygen and accelerates the progress toward reducing conditions, and such effects are difficult to generalize. Variable bottom currents might also effect scour and deposition to periodically expose and distribute sediment containing the radionuclides.

Each site will have to be evaluated on its own merit as regards capabilities of containment of radionuclides. Certainly at the 2800 meter site, a low energy depositional environment is apparent, however, it is not known as yet if burial and confinment will prevail before short circuiting effects results. The proximity of the 3800 meter site to the thalweg of the Hudson Submarine Canyon and the strong bottom currents in this area poses a threat for many drums while slumping and rapid deposition may accelerate burial of other drums.

SUMMARY

The characterization of the fine grained sediment at the Atlantic nuclear waste dumpsites provides basic data relating to the potential of the sediment to retain radionuclides that might be released from the waste drums. The high cation exchange capacity of the sediment and presence of clay minerals with high cation exchange capacities holds potential for this argillaceous material to act as a barrier to the release of radioactive material to the biosphere. However, to fully predict how well this retention might be effected , the sediment should be assessed by laboratory column infiltration studies of ocean sediment using in situ pH and Eh parameters and the existing environmental condition of the sediment at the site. Short circuiting mechanisms, such as organic complexing and biotubation, should be considered as possible release mechanisms in the sediment. Geologic stability and hydrodynamic considerations should also be evaluated in the barrier potential of the sediment to retain the radionuclides until they decay to innocuous levels.

BIOGRAPHICAL REFERENCES

(1) Rawson, M.D. and Ryan, W.B.F., Geological observation of deepwater radioactive waste dumpsite - 106 , U.S. Environmental Protection Agency Report 520/9-78-001, in press

(2) Hanselman, D.H., and Ryan, W.B.F., 1978 Atlantic 3800 meter radioactive waste disposal site survey - Sedimentary micromorphologic and geophysical analyses , U.S. Environmental Protection Agency Report 68-01-4836, in press.

(3) U.S. Army Corps of Engineers, Laboratory soils testing, EM-110-2-1906, Appendix II, Washington, D.C., (1970).

(4) Biscaye, P.E., Mineralogy and sedimentation of recent deep-sea clay in the Atlantic Ocean and adjacent seas and oceans, Geol. Soc. Amer. Bull., v. 76, p. 803-832, (1965).

(5) Biscaye, P.E., Distinction between kaolinite and chlorite recent sediments by x-ray diffraction, Amer. Mineralogist, v. 49, p. 1281-1289, (1964).

(6) Sayles, F.L., and Mangelsdorf, D.C., The equilibrium of clay minerals with seawater: exchange reactions, Geochem., Cosmochim. Acta, v. 41, p. 951-960, (1977).

(7) Shepard, F.P., Nomenclature based on sand-silt-clay ratios, Jour. of Sed. Petrology, v. 24, p. 151-158, (1954).

(8) Friedman, G.M., On sorting, sorting coefficients, and the log normality of the grain-size distribution of sandstones, Jour. Geology, v. 70, p. 737-752, (1962).

(9) Turekian, K.K., Oceans,Prentice Hall, Inc. Englewood Cliffs, New Jersey, 120 p.,(1971).

(10) Ames, L.L. and Rai. D., Radionuclide interactions with soil and rock media, U.S. EPA Contract 68-03-2514, v. 1, p. 3-168 - 3-203 , (1979).

(11) Emery, K.O. , and Uchupi, E., Western North Atlantic Ocean: topography, rock, structure, water, life and sediments, Amer. Assoc. Petroleum Geologists Memoir. 17, 532 p.,(1972).

(12) Hathaway, J.C., Regional clay mineral facies in the estuaries and continental margin of the United States east coast; in Nelson, editor, Environmental framework of Coastal Plain estuaries, Geol. Soc. of America Memoir 133,p. 331-358,(1971).

(13) Bondietti, E.A. and Francis, C.W., Chemistry of Technetium and Neptunium in contact with unweathered igneous rock ; science underlying radioactive waste management. Materials Research Science Meeting, Boston, Mass. 28 Nov - 1 Dec,p. 57 (1978).

(14) Heath, R.C., Barriers to radioactive waste migration, Oceanus, v. 20, p. 26-30 ,(1977).

(15) Seitz, M.G., et al, Migratory properties of some nuclear waste elements in geologic media, Nuclear Technology, v. 44, July (1979).

THERMAL AND CHEMICAL PROPERTIES OF SATURATED MARINE CLAYS IN A NEAR FIELD ENVIRONMENT*

J. L. Krumhansl and G. R. Hadley
Sandia Laboratories**
Albuquerque, New Mexico
U.S.A. 87185

ABSTRACT

Thermal and geochemical investigations were made on water saturated marine clays for the purpose of determining whether such sediments would have properties favorable for the disposal of high level waste. Thermal conductivity and diffusivity were found to remain relatively constant between 25°C and 300°C. The chemistry of seawater-sediment mixtures, however, undergoes marked changes with heating. A distinctly acidic condition may be generated above 200°C and many trace elements are solubilized. Mineralogic changes are dominated by the formation of a magnesium rich smectite clay. These changes are, however, of sufficiently localized extent that a considerable barrier to radionuclide migration would remain between an emplaced waste canister and the ocean floor.

*This work was supported by the U.S. Department of Energy under contract DE-AC04-76DP00789.

**A U.S. Department of Energy facility.

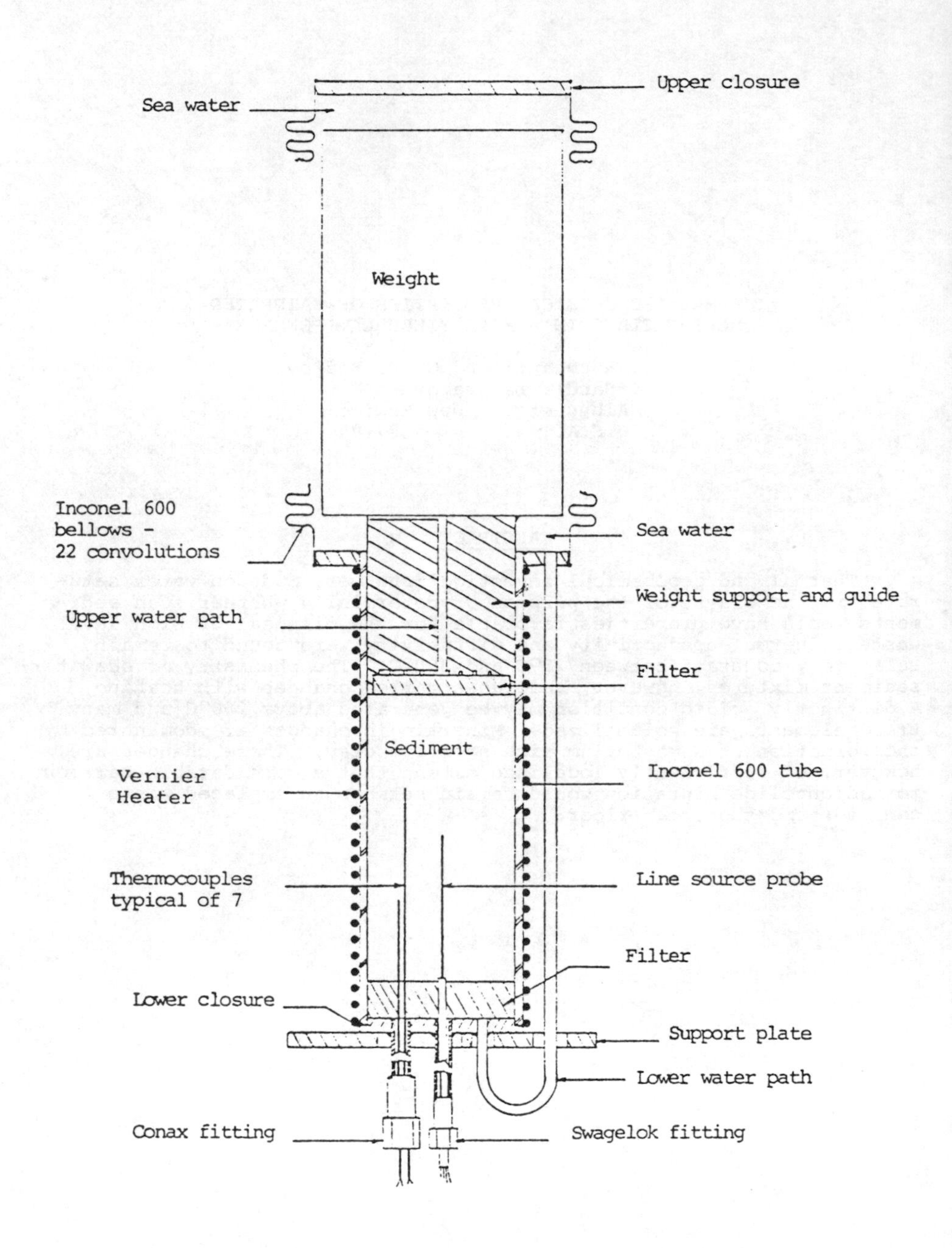

FIGURE 1. Schematic of Apparatus Used in Thermal Conductivity and Thermal Diffusivity Experiments

INTRODUCTION

Of the various argillaceous materials being considered for the disposal of nuclear wastes, abyssal marine sediments represent an extreme with respect to their water content. Depending on the mineralogy and age of the sample, water content may range from 80 to 250 weight percent of a sample's dry weight, with the higher values generally associated with smectite rich sediments characteristic of burial depths greater than 10.0 metres [1]. Because of their water content, such materials have a variety of properties attractive for the disposal of high level wastes. At ambient temperature, 1.8°C, when saturated with water such sediments have thermal conductivities of the magnitude required to dissipate heat from high level waste [2], while at abyssal depths boiling requires temperatures in excess of 400°C [3]. The high water content also results in the sediments having a plastic nature, hence it is highly unlikely that such material will have a natural joint pattern that could serve to short circuit the low permeability of the bulk sediment. When this factor is considered in conjunction with the highly favorable sorptive properties [4] of such clays it becomes apparent, as is demonstrated in a companion paper, to be presented at this symposium by McVey, Russo, and Gartling [5] that burial depths of but a few tens of metres may serve to safely isolate most radionuclides from the environment. This statement is, however, predicated on the assumption that the thermal and chemical processes in the immediate vicinity of the canister do not operate in such a manner as to compromise the far-field barrier. Programs were, therefore, initiated to characterize both the thermal and chemical behavior of sediments in the near-field environment.

THERMAL PROPERTIES MEASUREMENTS

The apparatus used in determining thermal conductivity and thermal diffusivity is illustrated in Fig. 1. Sediment was first packed around a line source and an array of thermocouples. A weight package was then placed on top of the sediment to simulate burial to a depth of 5 m. Once the sediment was compacted, a bellows was welded to the top of the apparatus and the entire assembly was lowered into a heater array inside a high pressure autoclave supplied by Battelle Columbus Laboratories. Thermal conductivity measurements were made using a line source probe method [6]. Once a sample had been heated to the desired test temperature, the probe was turned on and operated until a straight line was obtained on a graph of probe temperature versus the natural logarithm of time. The slope of this line is used to infer the thermal conductivity of the sediment. Thermal diffusivity was measured using a generalization of the well-known flash technique. The heater winding, wrapped directly around the sediment chamber, Fig. 1, was pulsed and the temperature probe then monitored. Determination of the arrival time of the thermal pulse at the probe allowed computation of thermal diffusivity.

Results for a sample of North Pacific illite, compacted until water constituted 89% of the sediment's dry weight, are shown in Figs. 2 and 3. Both conductivity and diffusivity remain constant below 300°C and then decrease slightly up to 420°C. Also shown are conductivity measurements made on cooldown. These measurements generally are higher than values measured during heating, probably as a result of compaction of the sediment during cooling.

A model for the thermal properties of saturated sediments has been developed and compared to the determinations illustrated in Figs. 2 and 3. The heat capacity per unit volume is easily modeled as a simple porosity weighting:

$$\rho c|\text{mixture} = n\rho c|\text{water} + (1 - n)\rho c|\text{mineral} \qquad (1)$$

where n is porosity and c is specific heat. For thermal conductivity, the preferred model is one in which heat is thought of as flowing in

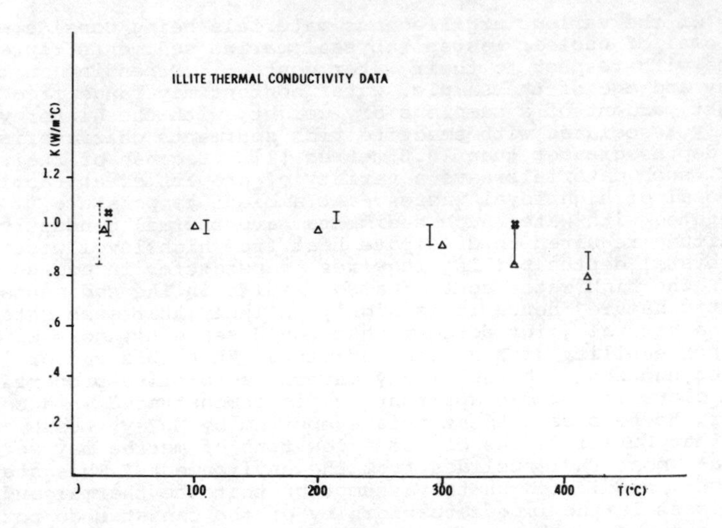

FIGURE 2. Thermal Conductivity of North Pacific Illite as a Function of Temperature: I Experimental Values During Heating, △ Theory (see text), I VonHerzen and Maxwell (1959), ✱ Experimental Values During Cooling

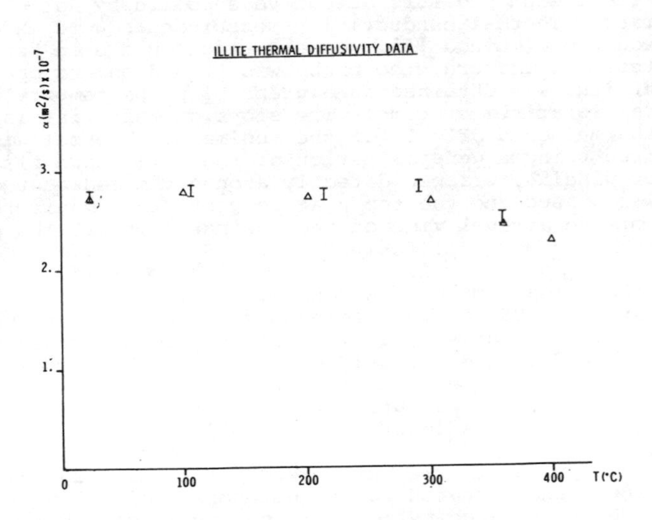

FIGURE 3. Thermal Diffusivity of North Pacific Illite as a Function of Temperature: I Experimental Values During Heating, △ Theory (see text)

parallel alternate slabs of water and mineral, present in proportions dictated by the porosity of the sample. Mathematically, this situation is represented by the following formula:

$$\kappa|mixture = n\kappa|water + (1 - n)\kappa|mineral \quad . \qquad (2)$$

Thermal diffusivity may then be arrived at by dividing the thermal conductivity, Eq. (2), by the heat capacity, Eq. (1):

$$\alpha|mixture = \frac{\kappa|mixture}{\rho c|mixture} \quad . \qquad (3)$$

Chemical Properties Measurements

From the thermal loadings presently anticipated for high level borosilicate waste and calculations such as those presented in a companion paper by McVey, et al., [5] it follows that temperature increases of several hundred degrees may be anticipated. The available literature indicates that temperature increases of this magnitude may induce changes both in the mineralogy of the silicate phases present, and in the composition of the pore fluid [7,8]. A joint program was, therefore, initiated between Sandia Laboratories and Dr. W. E. Seyfried at the University of Minnesota to investigate the chemical and mineralogical changes caused by heating various types of sediment: illite, smectite-rich, and carbonate ooze, all in contact with seawater.

From the literature, it is apparent that under hydrothermal conditions, seawater alone may become distinctly acid [8]. By virtue of the implications for both canister and waste form stability in the near-field environment, it was of concern to know whether sediment-seawater mixtures would behave in a similar manner. Fig. 4 summarizes the data obtained from water samples withdrawn from sediment-water mixtures at the stated temperature and then cooled to room temperature prior to the actual pH determination. For comparison, pure seawater (not in contact with sediment) that is filtered and quenched from 200°C and 50 MPa to room temperature has a pH of 6.2; when quenched from 300°C and 50 MPa a pH of 3.0 results. For seawater in contact with sediment there is an initial precipitous decrease in pH to values somewhat less than that expected for seawater alone. After a longer time has elapsed, the pH gradually rebounds as the silicate phases slowly react with the solution. The notable exception to this is the carbonate ooze experiment carried out at 300°C. In this case, the calcite present neutralized the acid as it was formed, thus preventing the pH from falling as far as occurred in the other 300°C runs. From these data, it follows that generation of an acidic condition may be mitigated in two ways--either by using a carbonate rich sediment or by keeping the peak temperature significantly below 300°C. Unfortunately, when carbonate rich sediments are heated to 300°C, relatively high concentrations of dissolved organics and carbon dioxide result, as well as there being a marked degradation in sediment texture. Seabed Program activities have, therefore, focused primarily on illitic and smectite rich sediments below the carbonate compensation depth, and assumed peak temperatures are significantly less than 300°C. The existence of a mildly acidic condition in immediate proximity to a hypothetical waste canister is assumed.

It was also found that metals such as Al, Fe, Mn, Zn, Ni, Cr, and Co, not found in normal ocean bottom seawater at 1.8°C in concentrations greater than 10 ppb, are leached from the sediment under hydrothermal conditions and achieve concentrations considerably in excess of 10 ppb, Table I. Their solubility can be attributed to the acidic condition at elevated temperatures. Their availability reflects a variety of exchange and mineralogic reactions occurring between the sediment and the seawater, as indicated by the marked changes in major component chemistry that also accompanied heating, Table I. Once in solution, such metals may diffuse or travel with

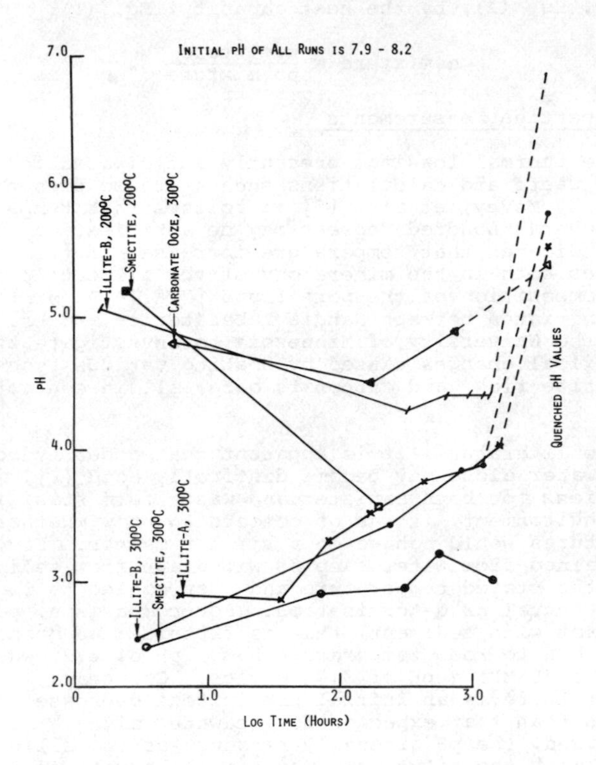

FIGURE 4. Quenched pH of Solutions in Contact with Illitic Sediment, Smectite-Rich Sediment and Carbonate Ooze (Water to Solid Ratio = 5)

convecting pore water until the acidic condition is neutralized, at which point sorption by clays or hydrous iron and manganese oxides serves to severely retard their further migration.

Since a wide spectrum of metals are mobilized under hydrothermal conditions, it is reasonable to assume that many of the metallic radioisotopes found in high level waste will behave in a similar fashion. The question arises, however, whether cooling of the canister would result in their removal from solution or whether they would remain continuously available for transport as assumed by McVey, et al., [5]. The entries marked "Q" in the temperature column of Table I are analyses of quenched solutions. That is, they summarize data obtained from solutions that were in contact with the sediment after the run had been cooled to room temperature. These values may, therefore, be taken to be representative of conditions that would exist in a closed system adjacent to a cooled waste canister. It was found that in most cases, an appreciable fraction of the metal solubilized at high temperatures remains in solution upon cooling, and in the case of strontium, the quenched values are consistently higher than the determinations made at elevated temperatures. Thus, mere cooling of the clay-seawater mixture will not result in all the solubilized metals being resorbed onto the clay, and, therefore, the assumptions made by McVey, et al., [5] concerning the possible long-time availability of radionuclides in the near-field environment appear to be valid.

Changes in the sediment itself are complex and strongly dependent on the temperature to which the sediment is heated. A few generalizations, however, may be arrived at by considering x-ray diffraction traces of the clays made at the conclusion of the hydrothermal tests, together with the changes in solution composition observed during and at the conclusion of the experiments, Table I. Generally, it is found that in the solution, magnesium decreases in concentration through the experiment while silica and potassium rise initially, level out, and may eventually start to decrease. It is inferred that these changes are in response to the precipitation, or formation by magnesium substitution, of a magnesium rich smectite clay. In the case of smectite rich clay samples, it was found that the illite interlayers were replaced to the extent that the sample experienced virtually complete recrystallization, Fig. 5a. The well crystallized illite in the illitic sediment is, however, more stable. In these sediments, it is apparent from Fig. 5b that the only changes that occur with heating are minor increases in the chlorite and smectite peaks. The iron-to-manganese ratio in these solutions suggests that the actual reaction involved is the conversion of a degraded iron rich chlorite to more stable magnesium rich chlorite and smectite.

To summarize, preliminary data pertaining both to the thermal properties and chemistry of sediment-water mixtures in the near-field environment have been reviewed. An illitic sample had a relatively constant thermal conductivity and thermal diffusivity up to 300°C. With respect to the initial premise regarding the apparent favorability of marine sediments for the disposal of high level wastes, it is pertinent to note that the thermal conductivity measured for the illite sample to 300°C is similar to values measured in the laboratory at room temperature for cores of marine sediment, [2] Fig. 2. It is not known whether smectite rich sediments will behave similarly, or whether their greater reactivity and water content will result in substantially different behavior under hydrothermal conditions. If a similar behavior is found, it will then be possible to calculate the peak temperature of a waste canister. This will, in turn, insure that convecting fluids will not transport radionuclides with sufficient speed to compromise the far-field barrier [5].

The second implication of the data presented here is that in the near-field environment, a wide spectrum of elements go into solution. In the event of fluid movement, such elements will remain relatively mobile until a sufficient volume of unheated sediment has been

Table I* Chemical Analyses of Filtered Water Samples Analyzed at 25°C After Being in Contact with Marine Sediments at 50 MPa and at the Stated Temperature

Sample Type	Time (hours)	Temp (°C)	pH	ΣC	Na	K	Ca	Mg	SiO_2	$SO_4^{=}$	Cl^-
			Major Components (Values in Parts Per Million)								
Illite A	1657	300°	4.0	649	11072	771	323	72	358	421	19511
	2351	Q	5.5	---	11143	616	1251	141	336	2522	19214
Illite B	1242	300°	3.9	649	12046	796	150	186	939	329	19253
	1247	Q	5.8	---	11377	321	820	197	481	2037	19634
Illite B	1414	200°	4.4	649	10928	729	221	683	460	1324	19148
	1437	Q	6.8	---	---	---	---	---	---	3056	19282
Carbonate	777	300°	4.9	3326	---	775	1453	8.0	754	---	19515
	876	Q	5.4	---	---	768	2209	34.59	429	---	19422
Smectite	1461	300°	3.0	422	---	1344	113	279			
Smectite	198	200°	3.6	275	---	913	140	1019	724	---	---
Starting Seawater**	---	25°	7.9-8.2	142	10763	399	412	1297	<0.1	2734	19375
Seawater**	---	200°	6.2	142	10760	399	120	1242	<0.1	1960	19300
Seawater**	---	300°	3.0	142	10760	399	41	1068	<0.1	1070	19300

*Seyfried and Janecke Unpublished Data
**Bischoff and Seyfried, Reference 8, no sediment present
Q Denotes water samples quenched to room temperature, in contact with sediment, filtered, and then analyzed.

Table I* (continued)

Minor Components
(Values in Parts Per Million)

Sample Type	Time (hours)	Temp (°C)	Fe	Mn	B	Al	Zn	Sr	Ni	Ba	Cr	Co
Illite A	1657	300°	40.2	226	18.7	0.23	11.1	7.9	---	1.41	0.015	0.027
	2351	Q	31.0	262	---	0.32	1.61	23.8	---	0.13	0.045	0.046
Illite B	1242	300°	7.1	669	19.2	0.30	16.11	2.81	0.010	0.27	0.012	0.08
	1247	Q	1.1	586	16.3	0.25	2.57	11.72	0.079	0.20	0.034	0.05
Illite B	1414	200°	--	355	17.0	0.1	0.78	2.7	0.018	0.10	0.009	0.03
	1437	Q	--	--	17.0	0.2	0.16	11.7	0.044	0.08	0.029	0.03
Carbonate Ooze	777	300°	1.6	2.6	--	0.4	0.68	64.3	---	>1.5	0.004	0.027
	876	Q	1.5	3.6	10.2	--	0.39	76.2	---	0.17	0.073	0.029
Smectite	1461	300°	<1.	104	--	0.16	19.3	--	0.02	--	--	0.11
Smectite	195	200°	0.3	48.6	--	--	1.6	3.9	---	--	--	--

*Seyfried and Janecke Unpublished Data
Q Denotes water samples quenched to room temperature, in contact with sediments, filtered and then analyzed.

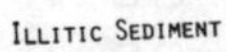

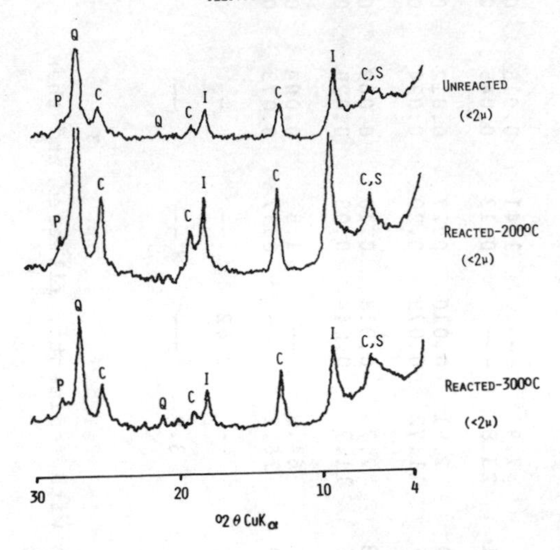

FIGURE 5a. X-Ray Diffraction Traces of Smectite-Rich Sediment Before and After Heating (Q = Quartz, P = Plagioclase, S-I = Mixed Layer Smectite-Illite, S = Smectite, I = Illite,

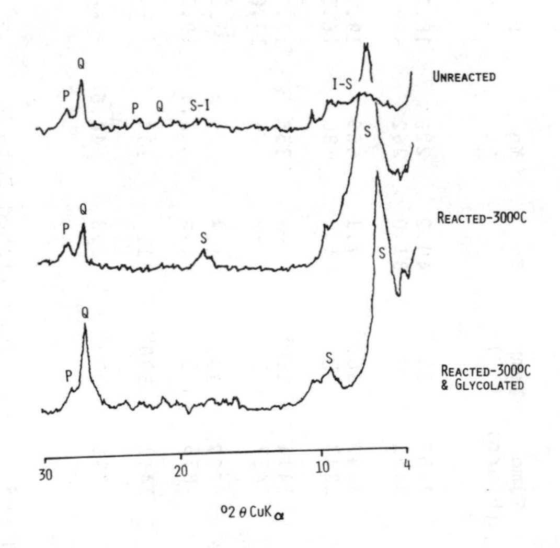

FIGURE 5b. X-Ray Diffraction Traces of Illitic Sediment Before and After Heating (Q = Quartz, P = Plagioclase, S-I = Mixed Layer Smectite-Illite, S = Smectite, I = Illite, C = Chlorite)

encountered to neutralize the acid generated during the thermal period. However, for a canister with a peak surface temperature of 300°C in illitic sediment, it can be calculated that less than two cubic metres of unheated sediment are required to accomplish this end. It follows that the several tens of metres of unheated sediment between the canister and the sediment-water interface would, therefore, still constitute a significant barrier to radionuclide migration.

The value of 300°C chosen in the above example is in excess of that presently thought prudent for the disposal of high level waste in a marine environment. Although the far-field barrier would be expected to retain its integrity even with these extreme conditions, it is unlikely that by virtue of the acid generated in solution, that it would be possible to guarantee the integrity of either a borosilicate glass waste form or a metal canister for as much as a century. Thus, for the multiple barrier concept to be fully operative, it is desirable to avoid distinctly acidic solutions, a condition that dictates that peak canister temperature remain significantly below 300°C.

REFERENCES

1. Silva, A. J., Oceanus, 20, No. 1, p. 31, 1977.

2. VonHerzen, R., and Maxwell, A. E., Journal of Geophysical Research, 61, No. 10, p. 1577, 1959.

3. Sourirajan, S., and Kennedy, G. C., Am. Jour. Sci., 260, p. 115, 1962.

4. Erickson, K. L., ACS Symposium Series 100, c15, p. 267, Sept. 15-17, 1978.

5. McVey, D. F., Gartling, D. K., and Russo, A. J., "Thermal/Fluid Modeling of the Response of Saturated Marine Red Clays to Emplacement of Nuclear Waste," NEA/OECD Conference on the Use of Argillaceous Materials for the Isolation of Radioactive Waste, Sept. 10-12, 1979, Paris, France.

6. Jaeger, J. C., Aust. Jour. Phys., 9, p. 167, 1956.

7. Helgeson, H. C., Brown, T. H., and Leeper, R. H., Handbook of Theoretical Activity Diagrams Depicting Chemical Equilibria in Geologic Systems Involving an Aqueous Phase at One atm and 0° to 300°C, 1969.

8. Bischoff, J. L., and Seyfried, W. E., Am. Jour. Sci., v. 278, p. 838, 1978.

Discussion

A.G. DUNCAN, United Kingdom

Can you estimate the size of experimental error which might arise from changes in the sediment due to decompression or recovery from the ocean bed ?

J.L. KRUMHANSL, United States

Some years ago Dr. Bischoff published a paper in Science saying that the pore fluids in cores recovered from abyssal marine environments undergo subtile changes when brought to shipboard conditions. Slight changes in K^+ and Mg^{++} concentrations were noted. These however are minute when compared with the chemical changes noted even in seawater alone that was brought to hydrothermal conditions. Thus the effects of heating the sediment to hydrothermal conditions vastly outweighs the consequences of transport to the ship from the ocean bottom.

D.F. McVEY, United States

We at Sandia know that it is important to investigate the migration characteristics of the different plutonium valence forms. We are sponsoring work to this end at Argonne National Laboratory for the seabed project. We expect results in about one year.

N.A. CHAPMAN, United Kingdom

I remain a little unconvinced about your deductions on nuclide migration related to metals remaining in solution during experimental quench procedure. Such conditions do not replicate the very slow cooling rate of a waste canister and the problem is essentially one of kinetics relating solution chemistry during slow cooling to convectional and diffusional nuclide migration. Any comments ?

J.L. KRUMHANSL, United States

Yes, it is a matter of kinetics, provided resorption of the radionuclides is not precluded by some aspect of the solution chemistry. In this case the solutions maintain a slightly acidic pH, a factor that could considerably interfere with radionuclide resorption until neutralization of the acid. A second factor which may alter the sorptive properties of the sediment are the mineralogic reactions that alter the sediment itself during the heating cycle. One might anticipate that clays would be better crystalized at the end of the treatment and hence have lower sorption capacities.

ADSORPTION ET ÉCHANGE CATIONIQUE AU COURS D'UN ÉCOULEMENT UNIDIRECTIONNEL EN MILIEU POREUX SATURÉ

J. ROCHON

B.R.G.M., Département M.G.A., ORLEANS, FRANCE

RESUME

Le présent manuscrit propose un certain nombre de modèles pour simuler le transfert d'une substance ionique interagissant avec une matrice poreuse homogène. Les types d'interaction étudiés sont l'adsorption, réversible ou non, et l'échange cationique, phénomènes prépondérants dans les milieux contenant de l'argile. Ces modèles ont été conçus avec le souci de simplification nécessaire pour leur utilisation ultérieure à des interprétations de données de terrain ou à des calculs prévisionnels.

SYMBOLES UTILISES

C : Concentration volumique du cation polluant en solution

C_o : Concentration volumique du cation polluant en solution à l'injection

D : Coefficient de dispersion longitudinale

d : Diamètre de la colonne de laboratoire

K : Constante d'équilibre de l'équation de LANGMUIR

K_d : Coefficient de distribution entre phases liquide et stationnaire à des concentrationstrès faibles

K_m : Constante de sélectivité de l'échange cationique

k : Constante cinétique de transfert entre phases

L : Longueur de la colonne de laboratoire

n : Porosité cinématique

Q : Débit volumique de l'eau

Q_o : Capacité d'échange cationique ou d'adsorption

S : Concentration volumique du cation polluant dans le solide

T : Somme des concentrations volumiques de tous les cations en solution

T_o : T avant injection du cation polluant

t : Temps

t_o : Temps de passage de l'eau $= \frac{L}{u}$

u : Vitesse interstitielle de l'eau

V_{inj} : Volume de cation polluant injecté

V_o : Volume des pores $= Qt_o$ à débit constant

x : Abscisse axiale

Σ_o : Concentration d'injection du cation polluant ramenée au volume des pores de la colonne $\Sigma_o = C_o \frac{V_{inj}}{V_o}$

INTRODUCTION

Un des moyens les plus efficaces pour assurer la sureté d'un stockage géologique de déchets radioactifs est d'assurer leur confinement sur les lieux mêmes où ils seront entreposés. Ceci met en exergue l'importance qu'il faut accorder aux barrières géochimiques naturelles existantes ou artificielles qui seront disposées entre les conteneurs de stockage et l'environnement humain. La barrière artificielle peut être parfaitement définie et peut, par conséquent, fournir à son niveau des renseignements précis sur le devenir des radioéléments en solution.

Le désir de prévision de la propagation des éléments dans le milieu nécessite de passer par des modèles susceptibles de simuler les mécanismes hydrodynamiques et physico-chimiques de migration. Le projet français de cette modélisation au niveau du massif est exposé dans [1]. De notre côté, nous nous sommes plus particulièrement intéressés au terme d'interaction éléments-minéraux et nous avons tenté une approche de la modélisation des phénomènes de transfert dans le milieu poreux homogène que pourrait constituer la barrière géochimique artificielle.

1 - CHOIX DES MODELES

Etant donnée la diversité des espèces susceptibles d'être contenues dans une eau de lixiviation de déchets radioactifs et la diversité des minéraux naturels, nous ne pensons pas qu'un seul modèle suffise pour simuler les migrations.

Le choix d'un modèle de migration pour tel ou tel élément et telle ou telle barrière nécessite de connaître le mieux possible :

- l'espèce aqueuse sous laquelle se trouve l'élément en solution
- la nature et la qualité de sa rétention par les minéraux constituant la barrière.

Ceci conduit à la conception d'un modèle spécifique à chaque couple élément-minéral.

Les modèles que nous présentons sont tous basés sur la même expression analytique de l'hydrodynamique pour un écoulement unidirectionnel, isotherme, à débit constant, en milieu poreux saturé [2].

$$D \frac{\partial^2 C}{\partial x^2} - \frac{u \partial C}{\partial x} = \frac{\partial C}{\partial t} + \frac{1-n}{n} \frac{\partial S}{\partial t}$$

Ce qui différentie les modèles existants est la façon dont est exprimé le terme d'interaction solide-solution $\frac{\partial S}{\partial t}$.

Selon les phénomènes d'interaction prépondérants, mis en évidence par des essais en "batch" et en colonne de laboratoire, nous proposons trois modèles différents pour interpréter les restitutions de radio-éléments. Les types d'interaction considérés sont :

- la sorption irréversible
- la sorption réversible
- l'échange d'ions.

Selon le phénomène d'interaction envisagé, nous aurons donc à résoudre un système d'équations constitué par l'équation de dispersion (1) et par l'expression de $\frac{\partial S}{\partial t}$.

2 - LA MIGRATION AVEC SORPTION

2.1. Sorption irréversible (modèle SI)

Si le phénomène d'interaction se manifeste pour une consommation en solution de l'élément sans de retard détectable par rapport à l'eau, ceci peut correspondre par exemple à une fixation irréversible ou à une précipitation sans remise ultérieure en solution.

En supposant que la cinétique de transfert entre phase est linéaire le terme d'interaction s'écrit : $\frac{\partial S}{\partial t} = kC$.

Ce modèle est à un seul paramètre k.

2.2. Sorption réversible (modèles S.R.)

Si la rétention d'une substance sur un solide est une adsorption, son isotherme répond le plus souvent à une représentation de LANGMUIR.

$$S = \frac{KQ_0C}{1 + KC}$$

Un soluté soumis à une telle adsorption aura tendance à être retardé mais sera finalement entièrement restitué. Deux cas peuvent être envisagés :

- L'équilibre est atteint instantanément
- La cinétique est du deuxième ordre en sorption.

Les modèles correspondant sont à deux ou trois paramètres d'interaction.

Cinétique	Terme d'interaction	Paramètres
Instantanée	$\frac{\partial S}{\partial t} = \frac{KQ_0}{(1+KC)^2} \frac{\partial C}{\partial t}$	K, Q_0
Non instantanée	$\frac{\partial S}{\partial t} = k\left[(Q_0-S)\,C - \frac{S}{K}\right]$	k, Q_0, K

3 - MIGRATION AVEC ECHANGE D'IONS (modèles EI)

Les expressions analytiques des modèles précédents ne font intervenir que les concentrations (en phase liquide ou stationnaire) du cation dont on veut simuler la migration. En fait, comme le terme même l'implique, lorsqu'il y a "échange" il faut également tenir compte des autres cations existant dans le milieu.

Les modèles de migration avec échange que nous proposons reposent sur un certain nombre d'hypothèses concernant la chimie du milieu.

(H1) - Il n'y a pas de réactions secondaires, telles que neutralisation, complexation, précipitation, de sorte que les espèces ioniques sont conservées.

(H2) - L'échange est uniquement cationique c'est-àdire que les ions ou complexes neutres et anioniques ne pénètrent pas dans l'échangeur.

(H3) - Les coefficients de dispersion de tous les ions sont sensiblement les mêmes.

(H4) - Le milieu est homogène, isotrope et sa composition initiale est la même en tous points.

(H5) - L'ensemble de tous les cations en solution, autres que le cation considéré A^+ se comportent globalement comme un ion unique de même charge B^+ : $A^+ + \overline{B}^+ \rightleftarrows \overline{A}^+ + B^+$

La loi d'action de masse relative aux concentrations de l'échange entre le cation A^+ et les autres s'écrit :

$$K_m = \frac{[\overline{A}^+]}{[A^+]} \frac{[B^+]}{[\overline{B}^+]} = \frac{S}{C} \cdot \frac{T - C}{Q_o - S}$$

qui peut encore s'écrire sous la forme.

$$S = \frac{K_m Q_o C}{T + (K_m - 1)C}$$

La capacité d'échange Q_o est une constante.
Par contre la somme T des concentrations des cations en solution en un point à un temps donné dépend de l'état antérieur du système. En vertu de la relation d'électroneutralité, T variera comme la somme des concentrations des anions en solution. Les anions n'étant pas retenus par l'échangeur((H2)), T migrera sans être retenu, satisfaisant donc à la relation :

$$D \frac{\partial^2 T}{\partial x^2} - u \frac{\partial T}{\partial x} = \frac{\partial T}{\partial t}$$

Le système d'équations à résoudre pour simuler le transfert du cation considéré A^+ sera donc :

$$\left\{\begin{array}{l} D \dfrac{\partial^2 C}{\partial x^2} - u \dfrac{\partial C}{\partial x} = \dfrac{\partial C}{\partial t} + \dfrac{1-n}{n} \dfrac{\partial S}{\partial t} \\ D \dfrac{\partial^2 T}{\partial x^2} - u \dfrac{\partial T}{\partial x} = \dfrac{\partial T}{\partial t} \\ \dfrac{\partial S}{\partial t} = \ldots\ldots\ldots\ldots\ldots\ldots \end{array}\right.$$

Selon que la cinétique de transfert entre phases est, ou non, considérée comme instantanée, le terme d'interaction $\frac{\partial S}{\partial t}$ s'écrit :

Cinétique	Terme d'interaction	Paramètres
Instantanée	$\dfrac{\partial S}{\partial t} = \dfrac{K_m Q_o}{\left[T + (K_m - 1) C\right]^2} \left(T \dfrac{\partial C}{\partial t} - C \dfrac{\partial T}{\partial t}\right)$	K_m, Q_o
Non Instantanée	$\dfrac{\partial S}{\partial t} = k \left[(Q_o - S) C - \dfrac{S\ (T - C)}{K_m}\right]$	K_m, Q_o, k

Les modèles correspondant sont à 2 ou 3 paramètres d'interaction.

4 - METHODES DE RESOLUTIONS ET D'AJUSTEMENTS DE COURBES

4.1. Résolutions des systèmes d'équations

Les résolutions des systèmes ainsi définis ont toutes été faites par discrétisation en différence finies "centrales" pour le terme dispersif, "arrière" pour le terme convectif, en corrigeant également la diffusion numérique apportée par l'approximation implicite ou explicite. Les incréments d'espace et de temps ont été choisis de manière à assurer la stabilité du schéma numérique.

Les conditions aux limites utilisées ont été :

- Limite supérieure : injection à flux constant d'un créneau de concentration d'un ion préalablement étranger au système dans une colonne alimentée par une eau de composition donnée.

$$t < 0 \qquad S = C = 0 \qquad T = T_o$$

$$\left.\begin{array}{ll} 0 \leqslant t \leqslant t_{inj} \ ; & C_o \neq 0 \\ t > t_{inj} \ ; & C_o = 0 \end{array}\right| \quad \left\{\begin{array}{l} uC_o = \lim\limits_{x \to o} \left(uC - D \dfrac{\partial C}{\partial x}\right) \\ u(T_o + C_o) = \lim\limits_{x \to o} \left(uT - D \dfrac{\partial T}{\partial x}\right) \end{array}\right.$$

- Limite inférieure : la dispersion est négligée en sortie du système :

$$o = \lim_{x \to L}\left(\frac{\partial C}{\partial x}\right) = \lim_{x \to L}\left(\frac{\partial T}{\partial x}\right)$$

4.2. Méthodes d'ajustements

Le nombre total de paramètres ajustables de ces systèmes d'équations est trop important pour donner des résultats significatifs [3].

- 2 paramètres hydrodynamiques
- 1, 2 ou 3 paramètres physico-chimiques.

C'est pourquoi nous avons adopté une méthode d'ajustement en cascade en définissant séparément les paramètres hydrodynamiques par un essai de marquage de l'eau (iodure, phénol, méthanol...). Par ajustement des courbes de traçage à l'équation simple de l'hydrodynamique (équation (1)), nous définissons les paramètres dispersif D et convectif u (notons qu'en milieu poreux saturé homogène à débit constant la vitesse interstitielle u tient compte implicitement de la porosité cinématique n).

Nous pouvons alors ajuster les 1, 2 ou 3 paramètres physico-chimiques à partir des modèles d'interaction spécifique présentés précédemment.

Certains ajustements de paramètres de ces modèles à des courbes de restitution expérimentales ont déjà été présentés en [4] pour :

- modèle "S.I." pour le couple technétium-sidérite
- modèle "S.R." pour le couple césium-quartz.

Dans ce manuscrit nous présentons les courbes obtenues en faisant varier certains paramètres du modèle "E.I.".

5 - SENSIBILITE DES PARAMETRES PHYSICO-CHIMIQUES DU MODELE "E.I.".

Pour déterminer la sensibilité des paramètres physico-chimiques du modèle à cinétique instantanée, nous avons effectué les simulations avec des paramètres hydrodynamiques et des conditions opératoires constantes.

$u = 5\ cm.min^{-1}$ $\qquad D = 1\ cm^2.\ min^{-1}$

$n = 0,50$ $\qquad V_{inj} = 0,2\ ml$

$L = 20,5\ cm$ $\qquad d = 0,45\ cm$

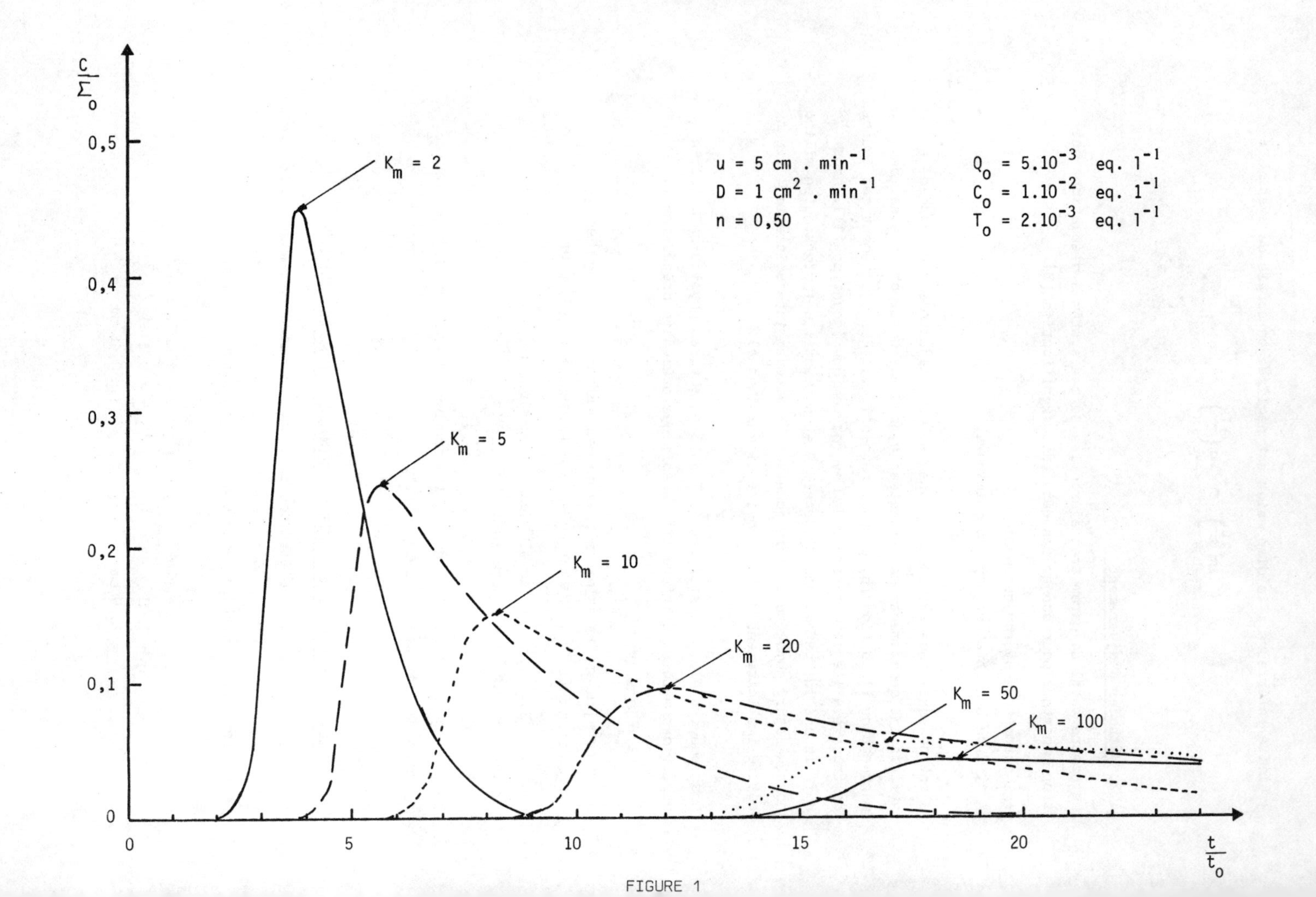
C/Σo
0,5
0,4
0,3
0,2
0,1
0
0
5
10
15
20
t/to
Km = 2
Km = 5
Km = 10
Km = 20
Km = 50
Km = 100
u = 5 cm . min-1
D = 1 cm2 . min-1
n = 0,50
Qo = 5.10-3 eq. l-1
Co = 1.10-2 eq. l-1
To = 2.10-3 eq. l-1

FIGURE 1

Nous avons choisi de présenter les résultats en concentrations et temps réduits.

$$\frac{C}{\Sigma_o} = f\left(\frac{t}{t_o}\right)$$

où t_o est le temps de passage, d'un soluté non retenu (temps de passage de l'eau) et Σ_o la quantité d'élément injecté ramenée au volume de pores.

$$\Sigma_o = \frac{C_o \cdot V_{inj}}{V_P}$$

5.1. Constante d'équilibre d'échange K_m

La figure 1 montre dans quelle mesure la constante de sélectivité K_m participe au retard et à l'étalement de la courbe de restitution d'un élément injecté à la concentration de $C_o = 10^{-2}$ eq/l. dans un milieu dont la capacité d'échange est de 0,2 meq/100 g et la densité de 2,5 g/ml. avec élution par une eau de concentration cationique à $T = 2 \cdot 10^{-3}$ eq/l

5.2. Capacité d'échange Q_o

Pour voir l'influence de la capacité d'échange sur la forme des courbes de restitution nous avons supposé que la constante de sélectivité était constante $K_m = 5$; les autres paramètres (C_o et T) étant les mêmes que précédemment. Nous avons fait varier Q_o de $2,5.10^{-3}$ à $2,5.10^{-2}$ eq/l., soit en supposant que la densité du support est de 2,5 g/ml de 0,1 à 1 meq/100 g (figure 2).

5.3. Concentration d'injection

Pour un même support et une même eau d'élution (K_m, Q_o et T constants), nous avons fait varier la concentration d'injection de l'élément. Les résultats sont présentés figure 3.

Nous voyons que les variations du pic de restitution (retard, atténuation, trainée) deviennent négligeables aux faibles concentrations d'injection dès que $T \gg (K_m - 1)C_o$, l'équation d'échange devenant linéaire.

$$S = \frac{K_m\ Q_o C}{T + (K_m - 1)C} \# \frac{K_m Q_o \cdot C}{T} = K_d \cdot C \qquad (2)$$

Ceci met en évidence les parts respectives du minéral (Q_o), de l'eau (T) et de l'échange (K_m) intervenant dans le coefficient K_d, celui-ci ne pouvant être pris en considération que si :

$$K_m \ll 1 + \frac{T}{C_o}$$

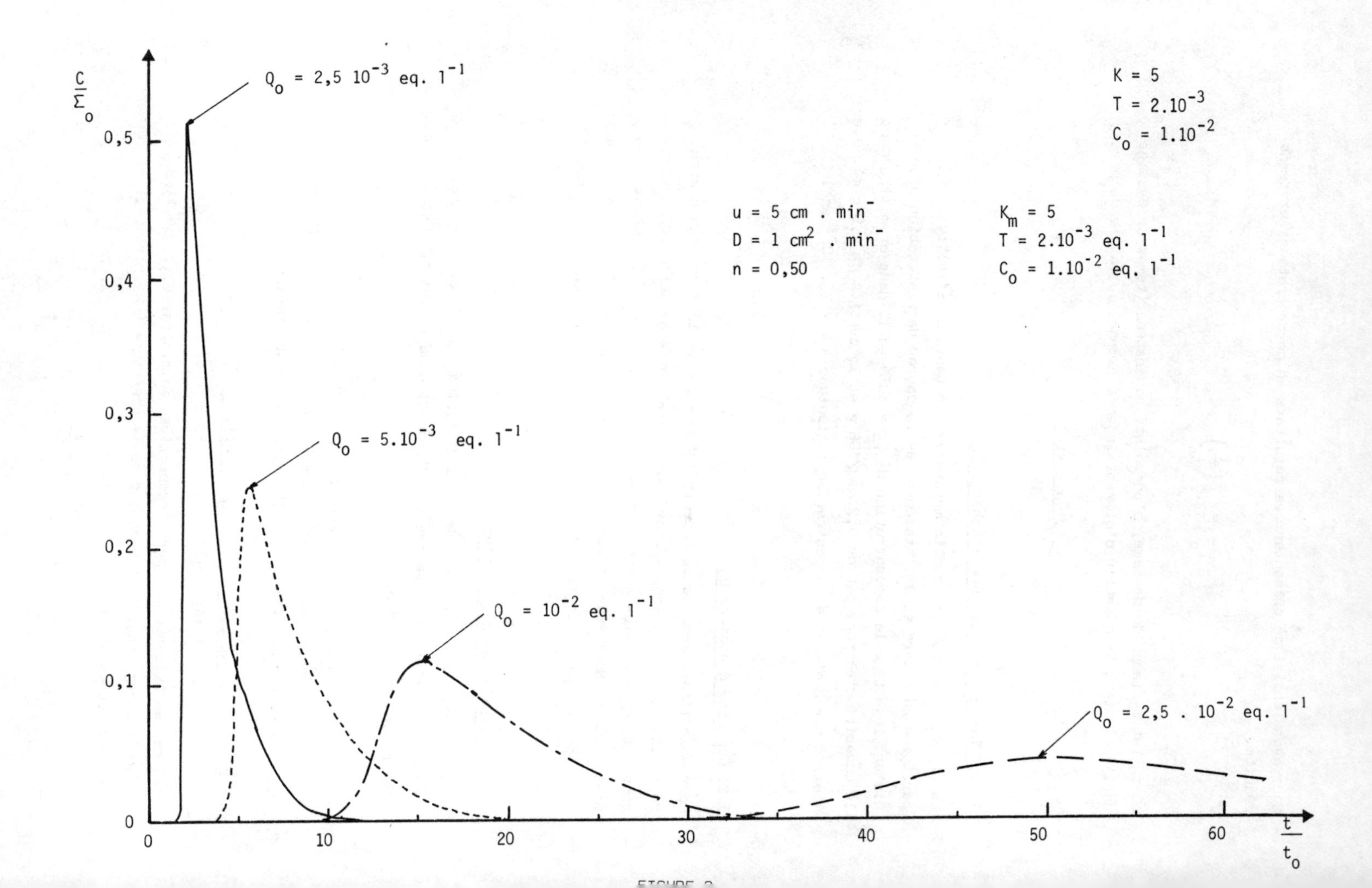
C/Σo
0,5
0,4
0,3
0,2
0,1
0
Qo = 2,5 10-3 eq. l-1
Qo = 5.10-3 eq. l-1
Qo = 10-2 eq. l-1
Qo = 2,5 . 10-2 eq. l-1
K = 5
T = 2.10-3
Co = 1.10-2
u = 5 cm . min-
D = 1 cm2 . min-
n = 0,50
Km = 5
T = 2.10-3 eq. l-1
Co = 1.10-2 eq. l-1
0
10
20
30
40
50
60
t/to

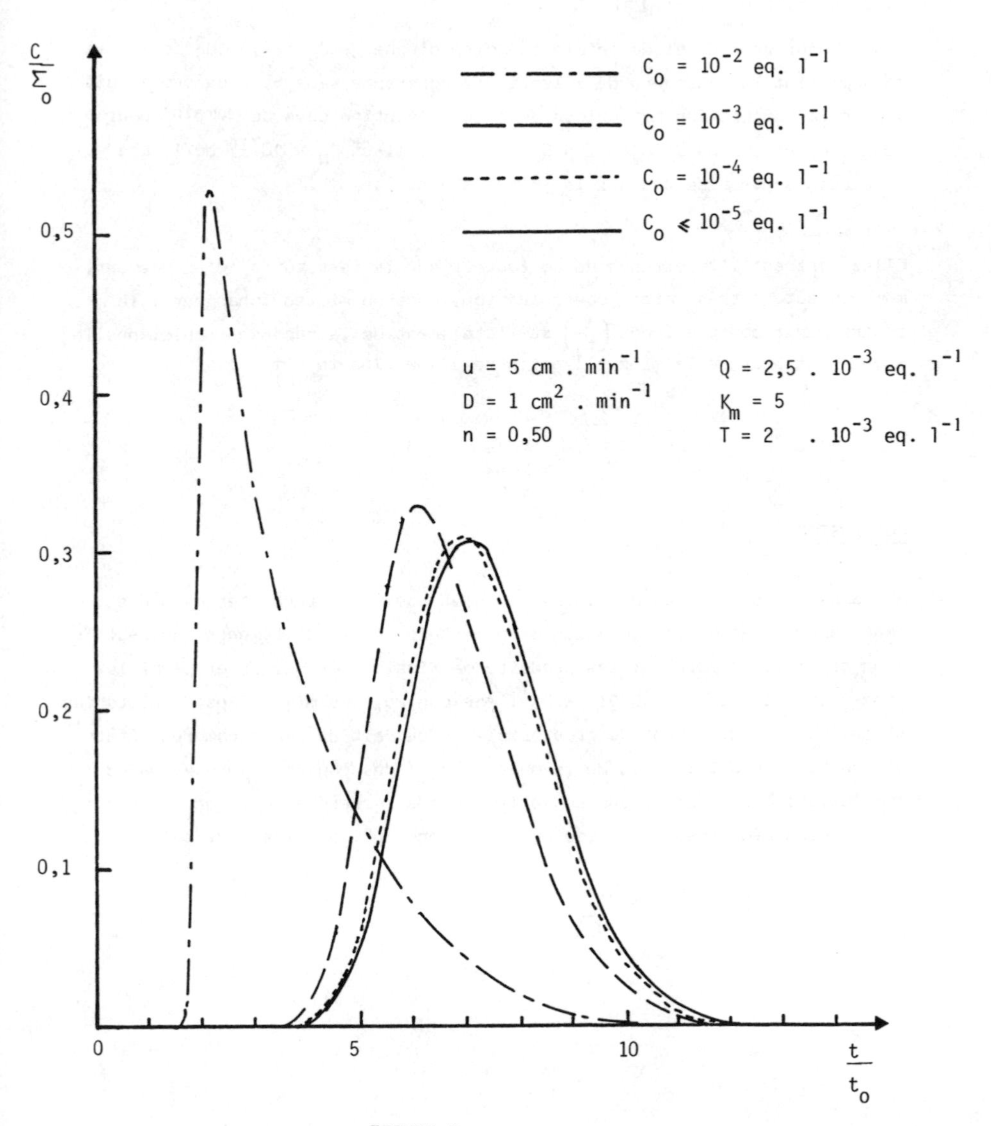

C/M_o
C_o = 10^-2 eq. l^-1
C_o = 10^-3 eq. l^-1
C_o = 10^-4 eq. l^-1
C_o ≤ 10^-5 eq. l^-1
0,5
0,4
0,3
0,2
0,1
u = 5 cm . min^-1
D = 1 cm^2 . min^-1
n = 0,50
Q = 2,5 . 10^-3 eq. l^-1
K_m = 5
T = 2 . 10^-3 eq. l^-1
0
5
10
t/t_o

FIGURE 3

Cette expression (2) montre également qu'aux faibles concentrations, le K_d d'un couple élément minéral, dépend de la composition cationique (T) de l'eau d'élution.

5.4. Composition cationique de l'eau éluante.

La plupart des eaux souterraines ont une minéralisation comprise entre 5.10^{-4} et 5.10^{-2} eq.l^{-1}. [5]

Aussi, pour un minéral de faible capacité d'échange $Q_o = 2,5\ 10^{-3}$ eq/l. et en supposant la constante de sélectivité constante $K_m = 5$, nous avons fait varier la composition cationique de l'eau d'élution dans ce domaine pour deux concentrations d'injection $C_o = 10^{-2}$ eq/l. et $C_o = 10^{-10}$ eq/l. Les résultats sont présentés sur les figures 4 et 5.

Elles montrent l'importance de ce facteur sur la restitution d'un élément, même injecté à très faible concentration, dans un milieu échangeur : le retard par rapport à l'eau $\left(\frac{t}{t_o}\right)$ et l'étalement de la réponse impulsionnelle augmentent lorsque la minéralisation de l'eau éluante diminue.

CONCLUSION

A l'aide de ces modèles de simulation nous pouvons interpréter avec une bonne approximation la migration d'un certain nombre d'éléments radioactifs dans des milieux poreux : les produits de fission Tc, Cs, Sr et parmi les transuraniens l'ion neptunyle NpO_2^+. Par contre, les migrations du plutonium et de l'américium posent de gros problèmes du fait de leurs nombreux états d'oxydation en solution et de la multiplicité des espèces hydrolysées ou complexées. Nous poursuivons actuellement nos recherches afin de concevoir des modèles réalistes pour simuler leur transfert dans les sous-sols.

BIBLIOGRAPHIE

[1] - BARBREAU A, BONNET M, GOBLET P., LEDOUX E., MARGAT J., de MARSILY G., PEAUDECERF P., SOUSSELIER Y.,
Premières évaluations des possibilités d'évacuation des déchets radioactifs dans les roches cristallines.
Colloque international sur l'évacuation des déchets radioactifs dans le sol, OTANIEMI, FINLANDE, 2 - 6 juillet 1979,
Rapport IAEA, SM-243/68

[2] - ROCHON J.,
Propagation de substances miscibles en interaction physico-chimique avec le substrat. Approche simplifiée pour l'utilisation en hydrogéologie.
Thèse de Docteur-Ingénieur en Chimie Minérale-Physique, Institut National Polytechnique de Grenoble, 25 janvier 1978.

[3] - VILLERMAUX J., ANTOINE B.,
Construction et ajustement des modèles mathématiques : une science ou un art ?
Bull. B.R.G.M. (2) III, n° 4 - 1978, pp. 327-339.

[4] - ROCHON J., RANÇON D., GOURMEL J.P.,
Recherche en laboratoire sur la rétention et le transfert de produits de fission et de transuraniens dans les milieux poreux.
Colloque international sur l'évacuation des déchets radioactifs dans le sol, OTANIEMI, FINLANDE, 2 - 6 juillet 1979,
Rapport I.A.E.A. -SM-243/155.

[5] - WHITE D.E., HEM J.D., WARING G.A.,
Data of geochemistry, 6ème Edition ; chapitre F : Chemical composition of subsurface waters. Geological Survey Professional paper 440-F.
United States Government Printing Office. WASHINGTON, 1963.

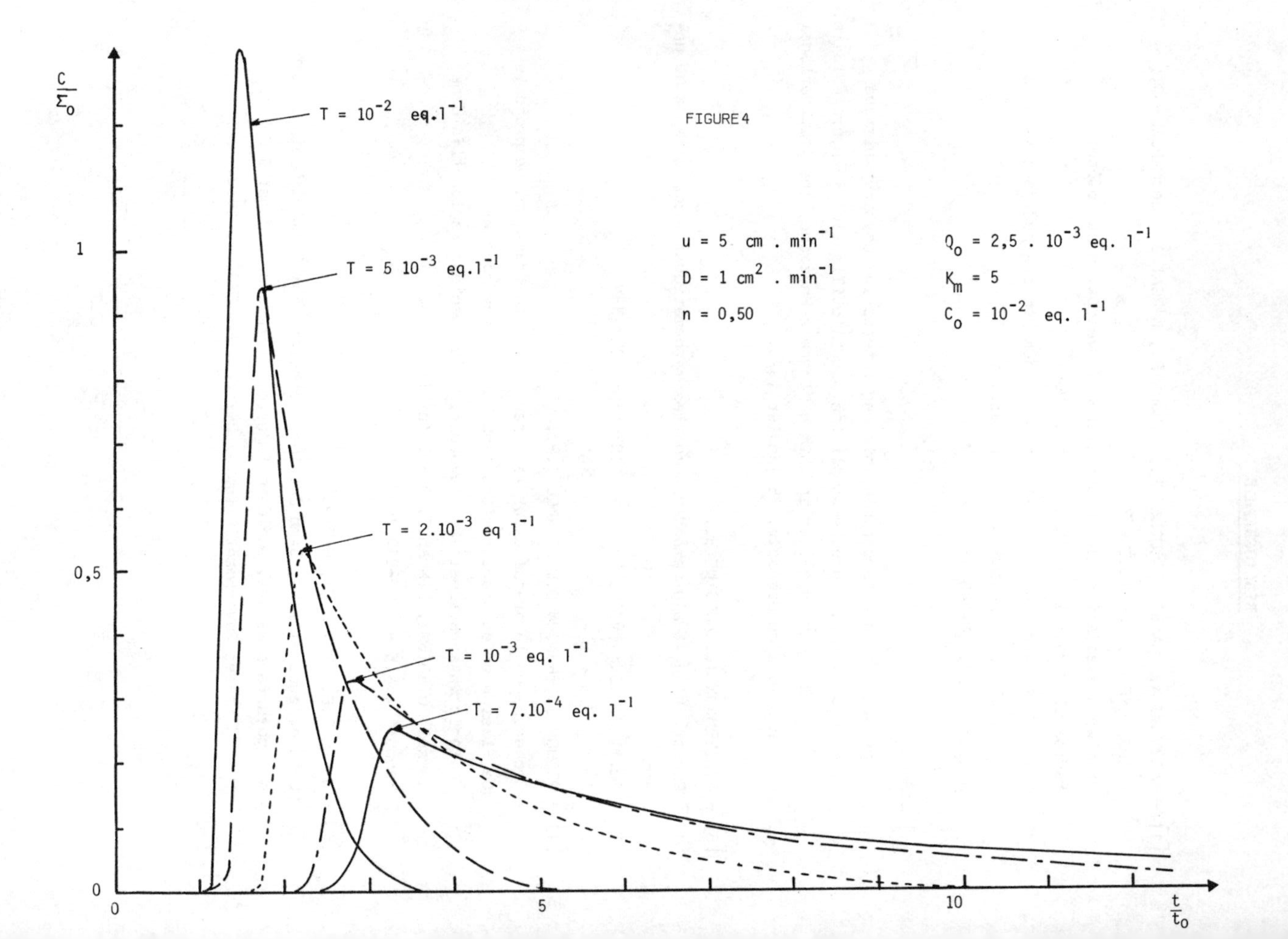
FIGURE 4
$\frac{C}{\Sigma_o}$
1
0,5
0
0
5
10
$\frac{t}{t_o}$
T = 10^{-2} eq.1^{-1}
T = 5 10^{-3} eq.1^{-1}
T = 2.10^{-3} eq 1^{-1}
T = 10^{-3} eq. 1^{-1}
T = 7.10^{-4} eq. 1^{-1}
u = 5 cm . min^{-1}
D = 1 cm^2 . min^{-1}
n = 0,50
Q_o = 2,5 . 10^{-3} eq. 1^{-1}
K_m = 5
C_o = 10^{-2} eq. 1^{-1}

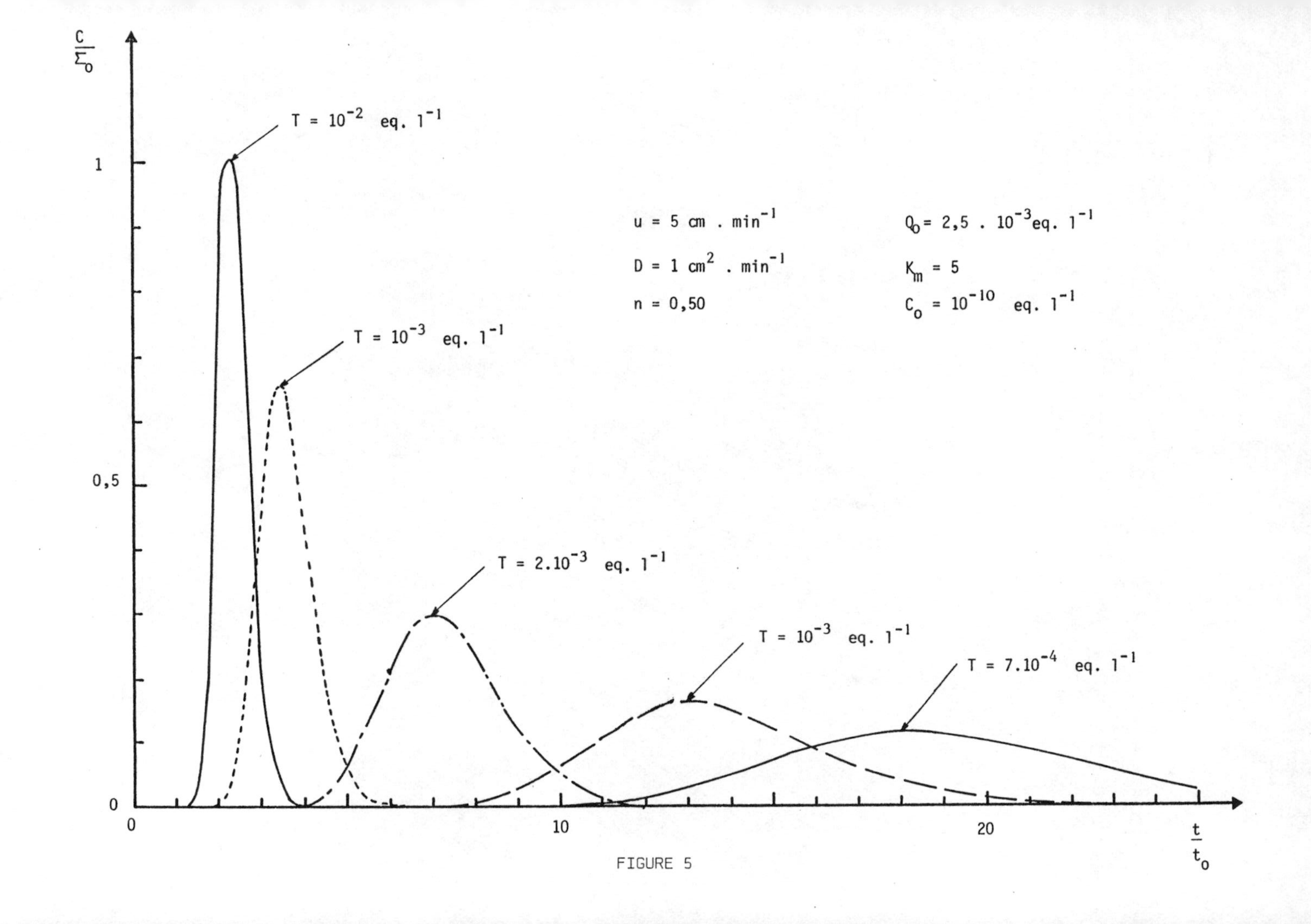
C/Σo
1
0,5
0
T = 10⁻² eq. l⁻¹
T = 10⁻³ eq. l⁻¹
T = 2.10⁻³ eq. l⁻¹
T = 10⁻³ eq. l⁻¹
T = 7.10⁻⁴ eq. l⁻¹
u = 5 cm . min⁻¹
D = 1 cm² . min⁻¹
n = 0,50
Qo = 2,5 . 10⁻³ eq. l⁻¹
Km = 5
Co = 10⁻¹⁰ eq. l⁻¹
0
10
20
t/to

FIGURE 5

GENERAL DISCUSSION — DISCUSSION GÉNÉRALE

N.A. CHAPMAN, United Kingdom

We can now move to a more general discussion of modelling studies both in general and specifically in relation to seabed disposal. There are various fields that we could profitably discuss and perhaps the one which would interest the modellers most is how we can introduce further elements of realism into the models which are currently being generated, which would help to compare one model against the other. Perhaps we could begin to hear from the modellers which are the most significant areas, where data is lacking at the present and how we should go about finding further information. Are there any comments on that particular aspect ?

D. DIRMIKIS, United Kingdom

As I mentioned in my own presentation we think it is necessary to take into account the creep behaviour of clay and there are no values at all, for that. At least we were not able to find any. We need values for example, for the spring constants and dashpots that I have shown and also for the activation energies that go with each of these properties.

A.R. LAPPIN, United States

If we consider the case of argillite, we cannot handle the fact that if joints do indeed open up above 100°C, they open up along pre-existing joints. Available models can handle, to a certain extent, axisymmetric joints which are generated during the process, but there is a big gulf between that and actually being able to understand how the preexisting systems of joints open up and what effect they will have on the properties of the rock. They certainly would destroy any axial symmetry in the system.

J. MARTI, United Kingdom

I think realistic modelling of the repository requires not only the coupling of the thermal and the stress/strain behaviour, but also the migration of pore fluids within the clay matrix. We have limited experience in doing coupled modelling of thermal and stress/strain behaviour or pore fluid and stress/strain behaviour of the matrix. We could, without too much conceptual difficulty, put everything together into a single model but there are a number of unknowns that we should resolve before doing that. Dr. Dirmikis has mentioned the uncertainties with respect to the creep behaviour of clay and the temperature dependence of the different material properties ; but we should keep in mind as well the difficulties we will likely have if we try to add to the model the hydraulic effects. The non-darcy flow, the existence of thresholds in the gradients that induce fluid flow need to be investigated. Unless we have experimental data and understand those processes and unless we can give mathematical formulations of their effects, there is very little we can do in the field of modelling.

D.F. McVEY, United States

I think that there is a lack of almost all sorts of data that are needed to make a model. I am referring to what Mr. Duncan brought up earlier in the day ; that is the need to do large scale in situ experiments. We are continuously confronted by the question : "Is what I measure in the laboratory, representative of actual repository conditions ?", at present we have no answer, at least for argillaceous sediments ; that is why I was glad to hear that the colleagues from Belgium are seriously considering getting down there to find out what goes on at depth. I think that the modelling task, although not easy, is perhaps a little easier than the experimental task. We can develop models ; given adequate computer capability we can model about anything, provided that we know the phenomena that we have to model and what the equations are. But in order to know that we have to carry out deep, in situ experiments. The point the previous speaker made about the coupling of the pore fluid pressure with the stress and the thermal behavior of the rock is very important. We have made some preliminary estimates on the basis of the experience with the Eleana argillite and with tuff. We have computed the pore pressure due to expansion of the fluid in the pores, using the permeabilities we think exist in situ. We see pressures developing in the pores that are significant in relation to the structural strength of the formation. It seems to me that there is a tremendous amount of work to be done and I think we will have to, I re-emphasize, get down where the waste is going to be placed in order to get the data that we need.

N.A. CHAPMAN, United Kingdom

One of the other areas, which we could profitably discuss is the connection between work which is being done on marine sediments and that which is being done for land disposal in clays and shales. For example we have seen how important near field effects appear to be in ocean sediment studies and perhaps we could hear some comments from people working on land disposal as to whether they feel these near field effects would be equally important in deep repositories in argillaceous formations.

J.L. KRUMHANSL, United States

In the last couple of years, I have been involved with the Conasauga near surface experiment and with the near field problems in marine sediments, this gives me some perspective on the nature of the problem. Marine sediments in the near field are probably more reactive than continental clays. By and large marine sediments are younger than most continental argillaceous sediments which are being considered for waste disposal. On the other hand there are certain near field effects in continental sediments which could be quite important. For example in the Conasauga test we found that, whereas normal Oak Ridge ground water has fairly low ionic strength, by the time we had completed the experiment the ionic strength had increased to the point where the water was saturated with calcium sulphate. After the experiment both sodium and potassium had increased and we observed in solution several ppm of strontium. This would say that the ionic strength perhaps had increased to the point where strontium was no longer well adsorbed by the clays. On the other hand the clay minerals themselves in the Conasauga test did not break down. There was no significant change in the clay mineralogy at the conclusions of the experiment in contrast to marine sediments. So I would say that probably there are near field effects in continental argillaceous formations which are important but they will not be the same effects that you see in marine sediments.

N.A. CHAPMAN, United Kingdom

Do we have any comments on this issue from those people who are actually involved in laboratory determinations of nuclide sorption ? Are the experiments satisfactory or there is a need for modifications ?

J. ROCHON, France

J'aimerais savoir s'il y a des essais de migration de radioéléments qui ont été faits avec l'eau de mer comme eau éluante. Quel Kd on obtient dans ce cas-là ?

J.L. KRUMHANSL, United States

The values which were reported on today are based on sea water and on the best available properties for marine sediments.

J. ROCHON, France

J'ai posé cette question, car il me semble que la force ionique de l'eau de mer est très importante et que les Kd que l'on pourrait mesurer sur des colonnes risquent d'être beaucoup plus faibles que des Kd mesurés, par exemple, avec une eau qui serait issue de massifs granitiques. A un fort Kd pourrait correspondre une restitution très rapide du radioélément, comme je l'ai précédemment montré dans mes courbes.

L.R. DOLE, United States

I think your question about the effect of ionic strength on Kd is addressed in the Kd programme which is conducted by Battelle Pacific Northwest Laboratory. One of the standard transport fluids in that study is a saturated brine solution, which represents an extreme, perhaps in case of elution from a salt deposit. In addition a systematic study has been conducted at Oak Ridge National Laboratory examining the effects of ionic strength on the Kd's of various minerals ; so there are indeed some studies that address the effects of ionic strength on Kd's.

J.B. LEWIS, United Kingdom

I am a bit perturbed about the mention of Kd here. There may well be several Kd's for the same element, for example in case of complexing solutions. What we need is not just an average value of Kd for a particular element but a determination of whether or not all of that particular radionuclide is adsorbed with that Kd value or whether a fraction is adsorbed with quite a different Kd, which might be the all important species.

A.G. DUNCAN, United Kingdom

I would like to make a general comparison between the deep ocean disposal programme and the underground disposal programme. It seems to me that the work on the effects of heat on the sorption characteristics of clay minerals is of much greater importance in the seabed than it is in underground formations ; if only by virtue of the fact that the emplacement technique for wastes in the seabed will leave us with a much smaller barrier. Having said that I wish to point out that we really do not know what sort of thermal transient the clay is going to see until we sort out the problem of the mechanical response of the sediments to the heat input. If convection were to take place the barrier would be disrupted ; then obviously the thermal transient would be much lower, but the result would be quite undesirable.

A.A. BONNE, Belgium

Je voudrais faire une remarque aux délégués de la Grande Bretagne. Je crois que pour les formations argileuses continentales et pour les formations d'argile des fonds océaniques on est en mesure de caractériser assez bien l'eau interstitielle du point de vue physique et chimique. On est déjà en mesure de préciser correctement tous les facteurs qui sont importants, comme la force ionique, les concentrations des ions solubles, etc. Ce qui fait que l'on peut déterminer les Kd dans une bande beaucoup plus étroite. Je ne vois pas quelle eau il faut prendre pour une argile appliquée comme tampon chimique dans un granite, tant qu'on n'a pas une idée de l'eau qui peut envahir un site de dépôt, car on peut envisager plusieurs types d'eau qui arrivent dans un granite. Tandis que pour les formations d'argiles continentales et pour le "seabed" on est en mesure de caractériser assez bien le type d'eau qui sera le milieu ambiant pour les déchets. Donc je crois que les valeurs de Kd pour les deux types de formations - argile continentale et "seabed" - peuvent être établies avec une certaine confiance.

J.B. LEWIS, United Kingdom

The point I was making is that if, for example, one takes plutonium, seldom it is in a single valence state. One normally has an equilibrium between valence states and also colloidal forms of plutonium ; I do not know what Kd values mean in this context.

D.F. McVEY, United States

There are some studies going on relative to the different valence states of plutonium in seabed sediments. These studies are being carried out at Argonne Laboratory. They are currently half to a third of the way through their experimental programme and I guess that some data should be available within the next year. I do not know enough about the preliminary results of their programme to summarise their findings.

J.L. KRUMHANSL, United States

Regarding the work at Argonne I know that they are discovering that it is indeed a very complicated system to work with. Every time they do something, they find out something new and interesting, but the generalities that would be useful for modellers have yet to emerge.

A.R. LAPPIN, United States

As regards the usability of Kd values and the ground water compositions in clay deposits, I am a little bit more pessimistic because I think that, for example, while we can characterise the ambient ground water, that has meaning only if we can show that there is not going to be any large inflow of water either through the mine workings or from an overlying aquifer. We cannot guarantee that the ambient ground water or the pore water in clays is going to remain unchanged until the time when radionuclides will migrate. As far as Kd values are concerned I am not sure what we can do with them unless the rock is completely free from joints and we can show that the porous flow approximation is valid. We have a lot of Kd values on argillite but we have not shown that they are valid for use in the field. The suspicion is that most fluid transport will take place through joints so we do not know what the effective surface areas might be ; maybe only a fraction of what you measure in the laboratory.

A.A. BONNE, Belgium

Je voudrais ajouter que, lorsque l'on fait des mesures de Kd dans nos laboratoires, on y ajoute toute une étude du comportement géochimique de l'élément, de complexation, de formation de colloïde, avant qu'on puisse interpréter le Kd. Ce que l'on faisait auparavant, je le dis peut-être un peu trop sévèrement, mais c'était un peu des essais aveugles, sans essayer d'interpréter ce qui se passe pendant l'essai.

N.A. CHAPMAN, United Kingdom

Well if nobody else has any comments or questions, I think we could profitably wind up here as time is getting on. It is clear that there is a lot more work that needs to be done. However it is reassuring to see how closely people involved in experimental studies and those involved in modelling are coming together. We have discussed again the perennial problem of the meaning of Kd measurements and perhaps one of the principle things we have agreed upon this afternoon is the need to add more elements of realism to the field experimental studies by taking them deeper underground. This in turn will add more realism to the models.

Session 5

Chairman-Président
Mr. T.F. LOMENICK
(United States)

Séance 5

T.F. LOMENICK, United States

This is the fifth and final session of the Workshop : a time for discussion. I think we are rather fortunate to have one morning to do nothing but discuss some of the problems which we would not have an opportunity to cover otherwise, perhaps we could even address some philosophical issues. Before we begin the general discussion, I would like to ask if there are any additional comments or questions related to the discussions in the four previous sessions.

R.H. HEREMANS, Belgium

Si vous me le permettez, je voudrais essayer en quelques minutes de vous faire part de la façon dont nous envisageons maintenant notre programme par rapport à l'étude sur l'interaction milieu géologique/déchets entreposés, et plus particulièrement en ce qui concerne la migration des radionucléides dans une masse argileuse.

Lorsqu'en 1975, nous avons pu procéder au premier forage de reconnaissance, nous étions heureux de pouvoir disposer d'échantillons en provenance d'un site de dépôt potentiel, prélevés à des profondeurs envisagées pour le stockage. Les chimistes ont procédé, sur ces échantillons, à une série d'analyses chimiques, minéralogiques, etc., les radiochimistes se sont lancés dans la détermination du pouvoir d'échange ionique de cette argile.

A l'époque, nous avions procédé d'une façon très classique. A partir des carottes de forage, nous avons prélevé des échantillons d'argiles, que nous avons mis en suspension dans une eau provenant de la nappe aquifère au-dessus de la formation d'argile et nous avons ajouté, sous forme de traceurs, les radionucléides dont nous voulions étudier l'absorption. Nous avons obtenu des résultats de Kd et nous avons examiné les variations de ce Kd en fonction de la concentration des radionucléides, en fonction du pH des solutions, en fonction du taux d'irradiation de l'argile - allant jusqu'à 10^{12} rad, et en fonction de l'échauffement. Nous avons donc obtenu toute une série de valeurs expérimentales de Kd qui n'étaient pas, dans un certain nombre de cas, des Kd réels, c'était des Kd apparents, car il y a eu échange ionique et également d'autres phénomènes chimiques. Ces valeurs de Kd seraient utilisables si l'on imaginait un accident majeur, c'est-à-dire là où de grandes masses d'eau seraient mélangées à l'argile et aux déchets entreposés.

Au fur et à mesure de l'avancement de ce programme, nous avons essayé d'imaginer d'une façon beaucoup plus réaliste ce qui en fait pouvait se passer dans des conditions, disons tout à fait normales, après la fermeture du site expérimental. Imaginons donc la formation argileuse à l'intérieur de laquelle se trouvent les déchets entreposés. Il n'y a plus de contact entre le dépôt souterrain et la surface, mais il y a contact intime entre les canisters de déchets et la formation argileuse. Que peut-il se passer à ce moment-là ? L'argile contenant de l'eau, cette eau peut sûrement provoquer de la corrosion. Nous avons donc commencé à examiner ce problème de corrosion dans l'argile, puisque ce serait le matériau du canister qui en premier lieu devrait être attaqué et complètement détruit avant que la migration des radionucléides puisse débuter.

Pour étudier cette corrosion, il était nécessaire d'avoir une idée aussi précise que possible de la composition chimique de l'eau interstitielle qui se trouve dans l'argile. Nous avons donc imaginé des techniques de laboratoire pour essayer de déterminer la composition chimique de cette eau. Nous pensons que nous avons relativement bien cerné le problème en travaillant de la façon suivante : on fait des mélanges ou des suspensions d'argile dans de l'eau distillée, dans des proportions croissantes ou décroissantes, puis on procède à l'analyse chimique de cette eau. On peut se faire une très bonne idée de la composition chimique de l'eau interstitielle dans cette argile en connaissant la teneur en eau interstitielle et par extrapolation des valeurs obtenues avec différents rapports argile/eau distillée. C'est une donnée extrêmement importante, non seulement pour les études de corrosion, mais également pour les études des réactions chimiques pouvant se passer dans le massif argileux avec les différents radionucléides qui y ont été déposés. Donc c'est un premier élément important dont nous disposons, et nous pensons que les données que nous avons actuellement sont très représentatives de la réalité.

Sur la base de ces informations, nous avons débuté un programme de corrosion tout d'abord en laboratoire, où nous étudions différents métaux, différents alliages, coupons purs, cordons de soudure, échantillons traités thermiquement ou soumis à des efforts mécaniques, puis une seconde série d'expériences in situ dans un massif argileux (une carrière actuellement à ciel ouvert).

Si nous revenons maintenant à la situation réelle : il y a des canisters dans le massif argileux, en cas de corrosion il y a passage des éléments de ces canisters, sous forme ionique, et diffusion dans l'argile. La première chose dont il faut tenir compte est que ces ions métalliques provenant du canister peuvent occuper des places libres normalement pour l'échange ionique dans l'argile. Ce sont donc des places qui sont occupées par des éléments chimiques et qui ne sont plus disponibles pour les radionucléides. Après corrosion complète de ce conteneur, on arrive à un "léchage" de la matrice de verre, et à un passage progressif des radionucléides dans le massif argileux. Ce passage, comment s'effectue-t-il ? Il y a incontestablement encore une fois, puisqu'il y a de l'eau interstitielle présente, passage sous forme ionique et migration de ces ions dans le massif argileux. C'est réellement un phénomène de diffusion qui s'opère et nous essayons actuellement depuis quelques mois, de simuler ce qui se passerait, en laboratoire.

Les premières expériences que nous avons faites dans ce domaine consistaient à prendre des petites carottes d'argile que nous mettions en contact avec un simple papier buvard imprégné d'une solution du radionucléide que nous voulions étudier. Nous avons ainsi fait des essais avec les quatre éléments classiques que nous utilisons dans nos laboratoires : le césium, le strontium, l'europium et le plutonium. Après avois mis en contact durant plusieurs semaines ce papier buvard avec le bouchon d'argile, nous avons procédé à un micro-carottage dans l'argile et à un "saucissonnage" de cette mini-carotte. Nous avons ainsi pu mesurer la profondeur de pénétration des différents radioéléments dans la carotte. A partir de là, nous avons pu calculer un coefficient de diffusion qui est, je pense, d'un ordre de grandeur de 10^{-6}, 10^{-7}, c'est-à-dire de 1 à 2 ordres de grandeur inférieurs dans le bon sens, à ce que l'on trouve dans la littérature, autrement dit la diffusion est moindre que ce qui généralement a été constaté sur des argiles prélevées en surface, puisque c'est surtout dans le domaine de la pédologie qu'un examen a déjà été fait précédemment sur d'autres ions.

Nous sommes conscients que ces expériences ne sont pas encore entièrement représentatives de la situation en sous-sol, car il manque un élément important qui est la pression. Nous avons donc développé ces derniers mois, à partir de cellules oedométriques classiques, un dispositif qui permet de remettre sous pression des échantillons d'argile et de mesurer la diffusion, ou du moins de procéder à des expériences de diffusion des radioéléments dans cette carotte mise sous pression, et, après, de procéder à un échantillonnage comme je viens de vous le décrire. Voilà, brièvement exposé, le type d'expériences que nous faisons actuellement dans ce domaine et qui, nous le pensons, nous rapproche des conditions réelles que l'on peut rencontrer dans le sous-sol.

C.R. WILSON, United States

I understand that you are now going into the field with a shaft to do additional work and most of the experiments that you have done so far have been in the laboratory. I am wondering in which areas you think that field experiments would be necessary to confirm laboratory results and what aspects of the actual in situ conditions you have not been able to successfully duplicate in the laboratory ?

R.H. HEREMANS, Belgium

Nous croyons actuellement que nous nous rapprochons de très près des conditions réelles du sous-sol. Bien entendu, nous n'avons pas la prétention d'être complets dans nos recherches, et il n'est absolument pas exclu que l'un ou l'autre élément nous échappe encore à l'heure actuelle.

Nous essayons maintenant de simuler la mise sous pression de l'argile et pourrons également, dans des cellules oedométriques adaptées, travailler à des températures variables, mais nous ne pourrons malheureusement pas, pour le moment, simuler l'irradiation gamma. Quoi qu'il en soit, même si les expériences peuvent être considérées comme représentatives, il nous faudra confirmer ces résultats en sous-sol. Comment exactement le ferons-nous, je ne puis vous répondre à l'heure actuelle, mais nous sommes persuédés qu'il sera possible d'effectuer des expériences de confirmation dans le sous-sol.

G. SCHMID, Switzerland

Votre explication porte sur les radionucléides qui arrivent dans le sous-sol argileux par diffusion, mais une fois qu'ils sont dans l'argile, est-ce qu'il y a aussi un transport par dispersion ? Autrement dit, est-ce qu'il y a une circulation d'eau dans ces argiles ?

R.H. HEREMANS, Belgium

Le fait que nous connaissions avec une très bonne approximation la composition chimique de l'eau interstitielle, nous permet d'étudier les réactions chimiques qui peuvent se produire dans ce massif argileux entre l'eau interstitielle et les radionucléides qui se trouvent sous forme d'ions et qui passent dans ce massif argileux. Il y a incontestablement, cela a été vérifié, des réactions de précipitations qui peuvent se faire, par exemple entre le strontium qui arrive dans le massif et les ions sulfates présents. A partir de là, vous formez du sulfate de strontium qui peut se présenter sous forme solide et qui peut être entraîné éventuellement par une circulation d'eau dans ce massif argileux.

Je dois revenir maintenant à d'autres expériences que nous avons faites, et qui nous ont permis de déterminer la perméabilité de ce massif. Cette perméabilité est faible, nous n'avons pas pu démontrer jusqu'à présent qu'il n'y a pas de circulation d'eau dans le massif, mais nous n'avons pas pu démontrer non plus qu'il y avait effectivement circulation d'eau. Nous avons une série de résultats de laboratoires qui semblent nous indiquer que si il y a circulation, elle doit être extrêmement faible. Je ne veux pas entrer dans les détails, mais je veux simplement dire ceci, si l'on compare la composition chimique de l'eau qui se trouve dans les nappes aquifères au-dessus et en-dessous de l'argile, et connaissant les concentrations des sels présents dans l'argile, s'il y avait vraiment lessivage du massif argileux, et compte tenu des millions d'années qui se sont passées, théoriquement il serait impossible de retrouver certains sels dans le massif argileux. C'est donc une indication que la circulation d'eau est extrêmement faible. Quoi qu'il en soit, les expériences de diffusion que nous avons prévues maintenant en laboratoire, doivent, je pense, nous permettre de mieux approcher la réalité des choses et si ce n'est pas une réponse définitive, c'est au moins une réponse partielle à la question que vous avez posée.

B. FEUGA, France

Dans son exposé sur l'expérience du chauffage in situ réalisé dans les schistes de Conasauga, M. Krumhansl a indiqué qu'il avait observé que la perméabilité du terrain mesurée après l'essai était plus petite que celle mesurée avant. Il a cité quelques explications possibles de ce phénomène.

Je voudrais, pour ma part, présenter brièvement les résultats d'expériences de laboratoire qui ont été réalisées il y a trois ans par le BRGM et qui pourraient peut-être apporter un autre élément d'explication.

Ces expériences ont porté sur le comportement de silts très fins (à dominante quartzeuse) sous l'effet de cycles d'augmentation et de diminution de la température, entre 20° et 80°.

Les échantillons étaient placés dans un moule de type oedométrique, lui-même placé dans un bac que l'on pouvait remplir avec une eau à température contrôlée.

On laissait d'abord la consolidation se produire, sous des charges de 0,5 ou 1 bar.

Puis on augmentait la température par paliers, et on mesurait la déformation stabilisée à la fin de chaque palier.

La température était ensuite réduite, puis à nouveau augmentée, etc., suivant le même processus.

Les observations réalisées n'ont pas correspondu du tout à ce à quoi nous nous attendions au début. En effet, alors que des essais de même type, réalisés sur des sables fins, ne faisaient apparaître que le phénomène de dilatation thermique, avec les silts, ce sont des phénomènes de contraction qui sont apparus (cf. figure).

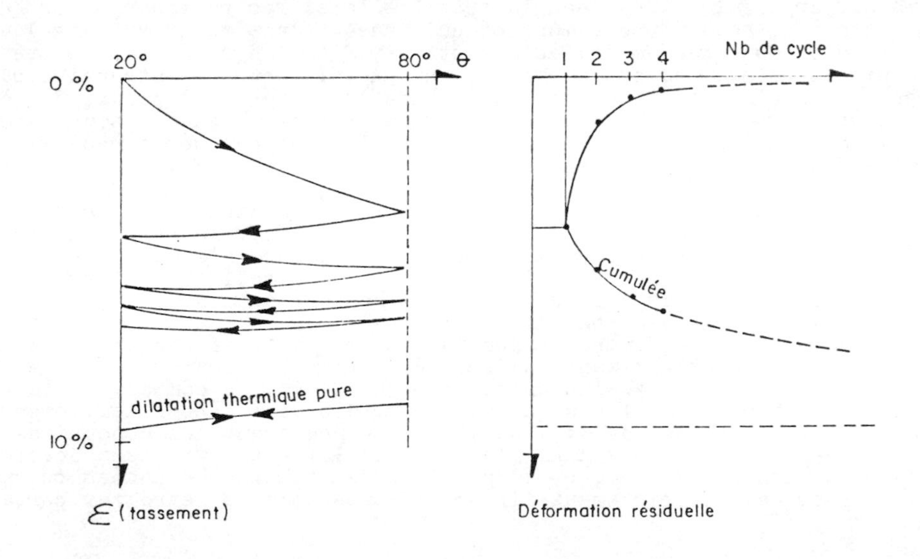

L'évolution subie au cours de chaque cycle de montée et descente en température n'était pas reversible, et à l'issue de chacun de ces cycles subsistait un tassement résiduel, qui décroissait avec le nombre de cycles.

Nous avons expliqué ce phénomène de la façon suivante : quand un corps dont les limites sont libres de se déplacer subit une augmentation de température, il se dilate sans apparition de contraintes nouvelles. Mais, si il est confiné, ce qui était le cas dans notre expérience, et ce qui est aussi le cas en un point situé au sein d'une formation géologique exempte de vides importants, l'augmentation de température se traduit par l'apparition de contraintes. Dans notre expérience, ces contraintes, s'exerçant aux points de contact des grains silteux les uns avec les autres, ont provoqué une plastification de ces points de contact, et donc un rapprochement des grains et un tassement du squelette solide.

Pendant les premiers cycles thermiques, les déplacements dus à ce phénomène étaient très largement prépondérant par rapport à ceux dus à la dilatation thermique. Il est évident qu'un tel resserrement du squelette solide ne peut que se traduire par une diminution de perméabilité du matériau, telle que celle observée à Canasauga.

Les essais que nous avons effectués montrent que ce phénomène est d'autant plus marqué que la dimension des grains est plus petite.

Aussi, bien que ces essais aient porté sur des silts fins, je pense que l'on aurait observé des phénomènes analogues sur des argiles.

Le domaine de contrainte et de température des essais que nous avons réalisés et de l'expérimentation in situ du Conasauga étant très voisins, cela permet un rapprochement entre les résultats obtenus.

M. Krumhansl a mentionné l'existence à Conasauga d'un dispositif d'auscultation comportant entre autres des extensomètres. Ces extensomètres ont-ils permis de mesurer des tassements ou au contraire des dilatations ?

J'ai le sentiment que, du moins en ce qui concerne l'évolution des perméabilités d'un milieu argileux sous l'effet de la température, deux phénomènes au moins peuvent intervenir ; ces phénomènes ont des effets inverses. Il s'agit, d'une part, de la création de fissures sous l'effet de la déshydratation, qui doit augmenter la perméabilité, et d'autre part, du phénomène de "consolidation thermique" que j'ai décrit, qui la diminue. Selon la nature du matériau, et le domaine de température et de pression considéré, l'influence respective de ces deux phénomènes peut être très variable. Il reste certainement beaucoup de travail de laboratoire à réaliser pour comprendre les comportements qui peuvent en résulter.

J.L. KRUMHANSL, United States

That certainly is a possibility. I can make two comments on the basis of the extensometer data from which we saw, more times than not, a slight dilatation rather than compression. On the other hand in some cases we observed a compression so it is not entirely clear cut one way or the other. A mechanism, such as you proposed, is perhaps a viable explanation despite the extensometer data. In my second talk yesterday I described some thermal conductivity and thermal diffusivity experiments on unconsolidated marine clays and in that case we saw exactly what you described. The clays underwent compaction during the course of the heating test. At the conclusion of the test they were more compacted than at the start, so we have also observed that phenomenon in a marine environment.

J. MARTI, United Kingdom

I was wondering whether you have been able to ascertain whether those cumulative strengths were due to a consolidation process. Knowing the coefficient of thermal expansion of the clay one should be able to estimate the stresses and from consolidation data it should be possible to predict the cumulative strengths. If it was a consolidation process the delay times should also be related to the coefficient of consolidation of the materials. Have you tried anything in that respect ? I suspect that if you were to plot the cumulative strengths on a log scale versus the number of cycles you should get something approaching a straight line and the slope of that line should be related to the consolidation coefficient. Have you tried any of those analyses or is the correlation of cumulative strengths to a consolidation process, just a guess from the shape of the curve ?

B. FEUGA, France

On n'a tenté aucune interprétation par le calcul de ces essais, qui avaient en fait un objectif très limité, sans aucun rapport avec le stockage des déchets radioactifs. Tout ce que je peux dire c'est que l'on a observé que la pente des courbes, de la déformation en fonction de la température, était beaucoup plus élevée que la pente qu'on aurait eu avec seulement le phénomène de dilatation thermique qui aurait donc été le coefficient de dilatation thermique du matériau en question. Maintenant l'implication des phénomènes que vous citez me paraît certaine, il y aurait lieu de développer un calcul théorique plus approfondi sur ce phénomène.

J. ROCHON, France

Ne peut-on pas penser qu'il peut aussi s'agir d'un phénomène couplé de diffusion thermique et de précipitation de la calcite, dont la constante de solubilité diminue avec la température. C'est une autre hypothèse qu'on pourrait peut-être avancer.

J.L. KRUMHANSL, United States

I suppose this is a reasonable guess, but we do not know enough about the chemistry of the clay that was used in this experiment to figure out whether that would be possible.

J. ROCHON, France

Je crois en fait que l'étude est à faire ; d'une part, essayer de déterminer les coefficients de thermo-diffusion, par exemple du calcium dans les argiles, et d'autre part, de faire une étude fine de ces argiles en contact avec une plaque thermique. Je pense que les Belges, étant donné l'appareillage qu'ils ont actuellement, seraient à même de faire ce genre d'expérience, alors qu'ils ont à mon avis, tendance peut-être à négliger cet effet thermique.

A.A. BONNE, Belgium

Je voudrais demander à M. Feuga si c'était dans des conditions de confinement qu'ils ont étudié cet effet de dilation et de diminution de volume.

B. FEUGA, France

C'était une cellule de type oedométrique, qu'on a fabriqué tout spécialement pour ces expériences ; je ne sais pas si c'est là où vous voulez en venir, mais on a tracé des courbes sans correction de la dilatation du moule et en tenant compte de cette dilatation.

A.A. BONNE, Belgium

Je voudrais savoir s'il y avait contact avec une solution et si cette solution était une solution simulant l'eau interstitielle ou de l'eau distillée ?

B. FEUGA, France

L'échantillon était bien saturé, en contact avec une eau, mais je ne me rappelle pas la composition de cette eau. Cependant, je dois dire, que du point de vue minéralogique, il s'agissait essentiellement d'un "silt" quartzeux.

A.A. BONNE, Belgium

Pour le moment, je crois qu'il y a plutôt des effets de rétrodiffusion d'éléments de l'argile dans la cire englobante et il ne faut pas s'attendre à des précipitations, mais à des phénomènes de dissolution.

J. ROCHON, France

Je parlais de précipitation de la calcite, car c'est un des seuls produits dont la constante de solubilité diminue avec la température, alors que pour pratiquement tous les autres minéraux c'est le phénomène inverse qui se produit. C'est pour cela que je parlais de calcite.

F. GERA, NEA

I think that this hypothesis would be very easy to test because there are some clays that contain practically no calcite and therefore if you observe the same effect in some of these clays you can exclude this particular mechanism. M. Feuga, do you happen to know how much calcite was present in your silt ?

B. FEUGA, France

Less than one per cent.

J.L. KRUMHANSL, United States

In this respect I should mention that the marine clays which did exhibit this compaction with heating cycles are very low in calcite. In fact, calcite is probably absent as the samples were recovered from below the carbonate compensation depth ; so we observe this phenomenon in materials where there is probably no calcite precipitation. On cooling, however we may get some silica precipitation as the concentration of silica in pore water increases on heating of these clays.

D.F. McVEY, United States

I would like to discuss a little bit more our experiment on marine clays. The second experiment, the one that provided the compaction data, was completed about a week before we left to come here, so the results are preliminary and I have not seen the detailed data. I will attempt to describe what effects we have seen. Basically, in this experiment we have a confined sample of seabed clay, under a constant lythostatic pressure, in contact with sea water, which is pressurised to an equivalent pressure of about 600 bars. The lythostatic load is equivalent to burial under about 5 meters of sediments while the pore fluid pressure in the clay is equivalent to a 6000 meters high column of water. The sample is compacted by placing a series of weights on a slurry of sea water and sediment. This is described in some of the seabed reports published by Dr. Silva of the University of Rhode Island. It takes approximately three to six weeks to compact the illitic sediment to the appropriate density. We have a compaction theory, which predicts very well the compaction of this clay and gives good agreement with the permeabilities we have measured in the laboratory on samples of the clay. When we placed the sample in the autoclave, before the start of heating,we observed that, as the pressure was brought up in the autoclave to pressurise the pore water, the sample began to compact again. I do not know the amount of the additional compaction but it was not negligible. We then proceeded to heat the sample and again we noticed considerable compaction. A normal compaction theory, which allows the water to expand and to migrate through the pores of the sample, would predict, as you heat, an initial swelling of the clay and then a gradual decay of pore pressure as the system comes into equilibrium. We did not see that, in fact in order to explain the behavior of the sample on the basis of the compaction theory we should assume a three orders of magnitude increase in permeability of the clay. We do not fully understand this phenomenon at this time but I would like very much to get any existing information on similar work that has been carried out in France.

B. FEUGA, France

Ce travail avait été fait pour un client privé, néanmoins, il nous a donné l'autorisation de le communiquer lorsqu'il sera publié.

T.F. LOMENICK, United States

I think this discussion has been a good introduction to the general subject of testing. I feel that testing is one of the most important parts of any programme, regardless of the type of rock which is being investigated. I would like to say a few more words about the hot cell test that I mentioned previously. Such test might provide essential information on the important parameters of argillaceous rocks. A canister of waste or perhaps an encapsulated spent fuel element should be surrounded by compacted buffer material and emplaced in a block of host rock, for example clay. In such experimental set up it would be possible to inject representative fluids and observe such things as the effects of radiolysis on corrosion. What we need, in the testing of argillaceous rocks, is a rather dramatic experiment that would get the attention of not only the scientists but of the managers as well. One problem we have encountered in waste disposal programmes is lack of continuity. It is rather difficult in some cases to make a large number of small laboratory tests look convincing to the management people who are responsible for funding the programme. In addition, at least in the US, it is becoming very difficult to work in the field, because the public is so sensitive to any mention of radioactive waste.

Therefore the hot cell test that I have proposed might help to bridge the gap between the small laboratory and in situ tests that have been performed and the rather expensive, sophisticated in situ tests that will eventually have to be performed at representative depths. A good deal could be learnt about the behavior of the disposal system in proximity to the radiation source.

In the investigation of argillaceous sediments we have got off to a late start. We are several years behind the salt programme and a few years I think behind the crystalline rock programme. The tests in the Stripa mine have been a great stimulus to the crystalline rock programme. They have focussed a good deal of attention on crystalline rocks and of course we have learnt some very important things. We need something similar in the study of argillaceous formations. Perhaps the tests that our Belgian colleagues will carry out at Mol will be exactly that. Most of us should consider participating in the Belgian experiment, so that, all together, we can learn something about the important parameters for radioactive waste disposal in argillaceous formations.

To my knowledge the excavation of a shaft and an underground room has not been done before for a test aimed at investigating radioactive waste disposal. Of course at Stripa the existing mine was refurbished and we have done something similar in the US in some abandoned salt mines. But to my knowledge the Belgian experiment will be the first instance of a shaft sunk and an excavation made for the purpose of studying radioactive waste disposal in a geological formation.

D.F. McVEY, United States

On the basis of some corrosion data that have been generated at Sandia, it seems that a hot cell test aimed at studying the interactions between host rock and waste in the presence of radiation should be a first priority.

F. GERA, NEA

Since this is a technical meeting it would be very useful to hear your personal technical opinion on some of these ideas. In particular what is your reaction to this proposal about a hot cell experiment ? I believe that it might be a very useful experiment ; on the other hand it would be essential to confine the sample or we would have the usual doubts about the behavior of pore fluids and the representativeness of the system.

T.F. LOMENICK, United States

Finding out about the interactions between buffer material and canister would be the primary reason for conducting the test. Obviously it would be difficult to reconstruct entirely the host rock. The data that could be obtained about the geological medium would be a somewhat secondary result of the experiment.

A.G. DUNCAN, United Kingdom

I understand the need to study the corrosion of the canister and the interactions with the host rock material. Is the proposal for a hot cell test aimed at looking at the effect of superimposing radiation ? If that is the objective can we not obtain the needed information from laboratory studies ?

T.F. LOMENICK, United States

I do not think that the data on the effects of radiation are available, especially in combination with the thermal effects. In such a test it would also be interesting to inject fluids similar to those found in the host rock. In case of salt rock brine would be the liquid to study, but in case of argillaceous rocks there would be other fluids to be injected into the system. This would give us the opportunity to learn about the behaviour of the kinds of fluid that could be found not only in salt and argillaceous formations on land, but in marine sediments as well.

J.L. KRUMHANSL, United States

I would like to make two comments. First, it is my experience that fairly large experimental systems are advantageous, since they enable us to assess the effects of gradients in the samples. Small laboratory experiments do not generally permit to set up temperature gradients, concentration gradients, radiation flux gradients, whereas a test such as the one proposed by Dr. Lomenick would. It was clear in the Conasauga near surface test that gradients made things happen. Constant conditions do not cause migration or other kinetic processes. The second comment regards the desirability of confining the sample. I think this is, to a certain extent, a good idea. However, most of the repository conceptual designs I have seen indicate that during the operation of the repository the waste canisters will be within a few metres of an opening ; therefore, at least for the purpose of studying the behavior of the system in the operational phase, a facility capable of simulating the confinement provided by the overburden is not needed.

F. GERA, NEA

This is a good point ; but a conventional mine is only one of the possible design options for a repository in argillaceous rocks. I am surprised that being involved in the seabed program you assume that that is the only way ; in some clay formations the matrix of drill holes can be a viable alternative. In that case you would probably want to plug the holes as soon as possible and you would be in a truly confined environment. At this point, since the emplacement options are still open, it would be useful to have a clear idea of what happens both in the case of an entirely confined environment, and in the case where there is an escape route for moisture and radiolytic gases. Another point, that I would like to make,is that it would also be possible to actually test the formation material. Our Italian colleagues have taken a large block of clay to the laboratory and carried out heating experiments. Maybe Dr. Brondi remembers the dimensions of the block. It is certainly possible to take a large block of clay, enclose it in some kind of steel jacket and put the whole thing in a hot cell. Finally I think that spent fuel or actual waste might not be needed for the first series of experiments. You are only interested in investigating the added effects of radiation, therefore a cobalt or cesium source might provide a much simpler way to obtain the needed answers.

R.H. HEREMANS, Belgium

Effectivement, l'idée de M. Lomenick semble fort attrayante au départ, mais nous n'avons pas eu suffisamment de temps pour réfléchir ou pour discuter de cette affaire ; je ne suis pourtant pas convaincu que sa réalisation soit tellement aisée, particulièrement dans le cas de l'argile. Nous avons envisagé, il y a deux ou trois ans, d'exécuter une expérience de transfert de chaleur avec une source simulée à dimension réelle, c'est donc l'expérience que nous faisons actuellement dans une carrière d'argile. Nous avions imaginé de faire cette expérience à Mol, c'est-à-dire de transporter à Mol un bloc d'argile suffisamment grand pour pouvoir faire cette expérience, mais nous nous sommes heurtés à pas mal de difficultés. Finalement, le coût d'une telle expérience devenant énorme, nous avons tout simplement abandonné le projet avant même d'avoir complètement solutionné le problème. Actuellement, nous louons quelques mètres carrés dans cette carrière, nous avons régulièrement la distance à parcourir, mais ceci nous revient sûrement moins cher que d'essayer, d'abord de façonner un immense bloc d'argile (parce qu'il faudrait qu'il soit immense), de le maintenir intact, de le transporter intact, de l'installer, de le confiner (parce que je suis persuadé aussi qu'il faudrait le confiner). Donc je crois quand même que ce sera une expérience extrêmement coûteuse, extrêmement délicate, et finalement la remarque peut-être des opposants à l'énergie nucléaire sera toujours la même : vous avez fait là une très belle démonstration, mais ceci n'est pas tout à fait représentatif des conditions réelles, etc.

T.F. LOMENICK, United States

Certainly cost would be an important consideration in such a test ; a few minutes ago we were talking about the difficulties of going into the field and conducting deep in situ tests. Even though the costs for such a test would be great, it certainly would not be as great as for sinking a shaft and making an in situ test in an argillaceous deposit. However the main point is that now we would not be able to conduct an in situ test, because of the political situation. At this time we could postpone the in situ demonstration test until the political situation becomes favorable or we could go back to the laboratory and conduct small scale experiments. An alternative is to consider these hot cell tests which would perhaps give us the best of both worlds. We would be able to work without outside interference and, hopefully, we would simulate quite well the in situ conditions.

J.B. LEWIS, United Kingdom

I am an advocate of in situ tests if for no other reason, to convince the public who in the end must decide whether or not disposal will take place. Public opinion will require in situ tests, but, as far as canister corrosion rates are concerned, I expect them to be negative. We are looking for canisters that will last for a very long period of time and if corrosion effects were observable during a test the canisters would not meet the required criteria. We need to study corrosion phenomena in the laboratory and develop the capability to extrapolate very low corrosion rates over long periods of time.

A. BRONDI, Italy

Dr. Gera said something about our laboratory test. The results of the laboratory experiment were quite good ; we also had an in situ heating experiment with ambiguous results.

I would like to stress that speaking of geological confinement, some useful information may be provided by the observation of the natural heating that has taken place around some natural heaters, for example some small laccoliths that exist in Italy and elsewhere. In central Italy we have pliocenic clays and flyschoid clays, which, in proximity to some intrusive body, have been exposed to temperatures maybe as high as 500 to 800°C. Thermal metamorphism has caused changes of the rocks surrounding the intrusive body. Then there is the case of geothermal fields, that are quite common also in the United States, and can simulate the conditions in a waste repository, but on a much larger scale. They can provide some useful information on the large-scale and long-term effects of heating. We know that in geothermal fields water is in the vapour phase with pressures that, in some geothermal sites of Central Italy, are as high as 80 to 100 atmospheres.

G.D. BRUNTON, United States

In reply to Dr. Lewis I believe that if we wait until we get a long-lived canister we will never dispose of the waste. The materials to be used for that canister would probably be too expensive and too rare to consider burying them. One of the results of an in situ experiment might be that there is no need for a canister, since the rock ensures adequate containment. Such an experiment would be very valuable if it succeeded in putting to rest the need for multiple buriers and long-lived canisters.

F. GERA, NEA

Geologists do claim that in certain geological formations there is no need for canisters or artificial barriers because the geological medium is an adequate barrier, but on the other hand we must recognise that there are public acceptance difficulties and redundant containment barriers may help to overcome the opposition to geologic disposal concepts. In at least a particular case, a reliable canister is essential ; I refer to seabed disposal. In this concept it is important to have an artificial barrier during the initial period when the waste is characterized by high heat generation rate because of the doubts about the possibility of convection in the sediments and accelerated migration of radionuclides. It is therefore essential to prevent the contact between waste and sediments until the thermal gradient is essentially gone. I do not think that finding suitable canister materials is going to be so difficult. There are already some preliminary studies that indicate that such materials exist and that their cost is not prohibitive. Even without going into such sophisticated concepts like the Swedish copper canister, it is possible to find relatively inexpensive material that should last a thousand years or more.

Secondly I wish to give you more details about the large block of clay that the Italians brought into the laboratory. The block is a cube with 80 cm side and a weight of about 1 tonne. It has not been easy to get a cube of this size in one piece. Several previous attempts failed because the sample fractured. Finally when one block was obtained, it was placed in a steel box and covered with paraffin. If such a sample were to be used for experiments under confining pressure a strong enclosure would be necessary. The cost of this particular sample was about $ 2,000, not including transportation costs from the quarry to the laboratory, that, of course, would depend on the distance.

R.H. HEREMANS, Belgium

Actuellement nous avons en cours d'exécution un programme relativement important pour la détermination de certaines propriétés géomécaniques de l'argile pendant congélation, après congélation et de l'argile chauffée. Nous avons donc mis au point une technique de prélèvement d'échantillons en carrière. Ces activités sont en rapport avec l'étude conceptuelle que nous avons faite et avec le projet de creusement d'une galerie souterraine. Nous prélevons avec une scie à bois des blocs d'argile de 50 cm de côté. Evidemment, je suppose que la qualité de l'argile joue un rôle, mais nous parvenons à prendre des échantillons d'une façon extrêmement simple, et de maintenir ces échantillons parfaitement conservés.

D.F. McVEY, United States

I would like to comment very briefly on the first of Dr. Gera's points. We have found some materials that look very good for producing waste canisters. On the basis of the tests that have been run so far it seems possible to have long life expectancy at relatively modest cost. The problem is that we have not yet evaluated these materials in a mixture of heat, radiation, marine clays and sea water. It is possible that when we mix all the ingredients together we find out that there is a problem.

A.G. DUNCAN, United Kingdom

I would like to ask if there is any indication that radiation is going to increase significantly corrosion rates over and above the effects of heat, sediments and sea water ?

J.L. KRUMHANSL, United States

For one set of experiments the answer is definitely "yes" ; for another set is "maybe". We are carrying out the experimental work at Sandia with a gamma ray source and a series of small glass tubes. The heating from the gamma rays brings the temperature in the reaction tubes to between 15 and 18°C. The first preliminary runs used a variety of brines relevant to the WIPP project, sea water and deionized water. Several kinds of steel were tested. After about a month the mild steel experienced a little bit of corrosion, while some stainless steel was badly corroded. This caused considerable interest at Sandia and the experiments were repeated using a wider variety of metals and different compositions of brines. The second experiment lasted about two months. Corrosion of the stainless steel samples was much less than in the previous experiment, but a number of reaction tubes has exploded presumably due to the generation of gas. I admit that we do not know yet what is going on, but there is some evidence that,under certain conditions, radiation can make an important difference, at least in respect to the corrosion of stainless steel.

A.G. DUNCAN, United Kingdom

Can I just be clear about this point ? The corrosion is very much faster in presence of radiation than in a comparable experiment with the same conditions, but without radiation ?

J.L. KRUMHANSL, United States

That is correct, for one set of experiments. In the other set of experiments radiation was rather irrelevant. This suggests that we have a lot to learn about the whole problem.

N.A. CHAPMAN, United Kingdom

This effect may be due,in the case of steels,to some sort of hydrogen embrittlement, as a result of radiolysis of water. Have you any comment on that ?

J.L. KRUMHANSL, United States

This is a possibility, we do not have a good idea of what kind of gas was generated during the experiment. As I said before some of the tubes exploded during the course of the experiment. In one case a tube was open and the content analysed. Apparently nothing had changed. We do not know whether it is hydrogen embrittlement or the generation of chlorine compounds. Something was corroding the stainless steel ; that is all we can say at this time.

J. ROCHON, France

Je voudrais faire quelques commentaires, inspirés par la discussion qui vient d'avoir lieu, sur l'intérêt des essais en laboratoire ou in situ. Si on recense un peu toutes les sciences qui interviennent dans ces problèmes de l'évacuation des déchets, il y a grosso modo la géologie, l'ingéniérie, l'hydrogéologie, la thermique, la géochimie. On peut peut-être séparer en deux les études à faire et les regrouper en essais in situ et en essais en laboratoire, avec une charnière qui serait par exemple la thermique. Je pense qu'il serait intéressant d'étudier in situ les problèmes géologiques, hydrogéologiques et thermiques. Mais, par contre, les phénomènes géochimiques seraient beaucoup plus simples à étudier en laboratoire, où on pourrait les coupler avec les phénomènes thermiques. De plus, ces essais en laboratoire auraient l'avantage d'augmenter l'échelle de temps, par exemple on pourrait accélérer les débits d'eau et donc avoir une étude géochimique correspondant à un grand laps de temps sur le terrain.

N.A. CHAPMAN, United Kingdom

I agree with M. Rochon because I believe that in situ tests, whilst they are extremely valuable, in the final analysis can only be treated as empirical experiments. There are so many unknown variables that we need some sort of laboratory control. In case of corrosion studies we may feel that we understand the behaviour of the metal, but we understand very little about the behaviour of the surrounding material. The two ways of studying this problem are to do very detailed and closely controlled laboratory experiments and then to find out what actually happens in situ. Only then we can try to relate the two.

T.F. LOMENICK, United States

Are there any other comments on my proposal for a hot cell test ? At this time, it is no more than an idea. It has not been funded yet. I do have some indication that there is interest in this type of test and perhaps over the next year we will begin to do some preliminary work along this line. We have heard some pros and cons about the proposal and if some of you is interested in such a test it would be useful to get together at some later date and talk about it in more detail. If we do this test some of the other countries might want to participate in one way or another, for example by providing samples of argillaceous rocks or by contributing instrumentation for measuring some of the parameters.

R.H. HEREMANS, Belgium

Je voudrais vous demander, M. Lomenick, si vous avez déjà examiné ce problème en détail, et si vous avez déjà un plan de travail pour ce genre d'expérience ?

T.F. LOMENICK, United States

To date we have a proposal for a test where rock salt is the host medium. Since the experiment has not been funded yet we have not done the detailed planning ; that would be the next step.

There are many other laboratory scale tests that one might want to consider for discussion this morning. We might also continue the discussion of in situ experiments both near the surface and at representative repository depths. A deep in situ test, such as the one being planned by the Belgians, requires careful planning and great expenses. I am not aware of any other country considering an experiment of that type at this time. However in the area of small scale laboratory tests there must be some interesting tests that we have not yet considered.

L.R. DOLE, United States

Can we identify developmental needs for in situ testing in argillaceous rocks ? I know that difficulties have been encountered with the extensometers and with sealing both heater and observation holes. Can we list what is needed so that the next generation of in situ tests will be better ?

F. GERA, NEA

This is a very important point. In situ experiments so far have measured essentially temperature distributions. They give us some feeling for what happens when a heat source is placed in argillaceous materials. But we have not looked yet into one of the key problems, that is the behaviour of fluids. A few years ago, when I was working on disposal in argillaceous formations, I had concluded that a very interesting experiment would be to look at temperatures and fluid pressures at the same time. This turns out to be difficult because the required instrumentation for measuring fluid pressures in clays is not available. There is a clear need for the development of some kind of pressure gauge, that can be used in clays at fairly high temperatures.

A.R. LAPPIN, United States

We tried in the Eleana experiment. We put gas pressure gauges in instrumented holes and found out that the vapour pressure was effectively controlled by the coldest portion of the hole, so that we measured very low pressures. The real problem in relation to fluids and, by inference, radionuclides migration in the Eleana argillite is that the rock mass is jointed and any determination would depend on the fractures existing at that particular location. It would be risky to assume that the observation at one site were representative of the formation as a whole.

D.F. McVEY, United States

In the framework of the seabed programme we are planning to measure pore pressures in marine sediments. This is a fairly common determination in marine clays. Sandia laboratory is developing some devices for measuring pressures in clays in the seabed. This instrumentation is being developed in cooperation with the National Oceanographic and Atmospheric Administration, which has great experience in this field. We intend to emplace about three pressure probes in regions characterized by different temperatures so that the variation in pore fluids pressure can be correlated with the different temperatures in the sediments.

We are also working in the laboratory to develop techniques to measure pore fluid velocity. This would enable us to carry out laboratory experiments that could be used to validate the analytical codes ; that is the codes that are used to predict convection within the porous media. In a situation where the fluids velocity is very low, it cannot be determined by measuring temperatures, therefore you have to attempt to measure the velocity itself. There are several possibilities : modifications of hot water probes ; magnetic means, similar to those used to measure turbulent flows ; and alternating current methods. This work should result in the development of some probes that we hope to test in perhaps a year.

T.F. LOMENICK, United States

Any other comments ? I would like to make an observation. After the discussions of the past two days I have the impression that near surface in situ heating tests are going to be a thing of the past rather shortly. The US tests are approaching conclusion and other countries do not show much interest in this type of experiment. If that is indeed the case, we are left with the extremes ; that is laboratory tests and full scale in situ tests such as the one being condisered by the Belgians.

A.R. LAPPIN, United States

It seems to be true, but I think it is a great mistake. While near surface in situ tests are expensive in comparison to laboratory tests, they have provided a great deal of information on how varied the behaviour of argillaceous rocks may be. We still do not know how to extrapolate the data to greater depth, but the difference in behavior between Eleana argillite and Conasauga shale has been very instructive. The explanation is probably that in case of the Eleana the rock was strong enough to sustain open joints and still behave as a coherent mass after dehydration, whereas the Conasauga was not. By looking at the differences and similarities in those two tests, we have acquired a better feeling for what the critical parameters are. We certainly would instrument deep in situ tests in these two rocks differently now, on the basis of what we have learnt about their near-surface behaviour. If we had designed deep in situ tests without the near-surface tests we probably would have used the wrong instrumentation and obtained useless results. Your first in situ test is likely to give you some unpleasant surprises, it is therefore wise to perform it at shallow depth.

J. MARTI, United Kingdom

I would like to ask if Dr. McVey could describe the operation principle of the probe used to measure pore fluid pressures in deep sea sediments.

D.F. McVEY, United States

I do not know enough details of the gauge at this time to answer your question. I would refer you to Roger Anderson, Lamont-Doherty Geological Observatory, Palisades, N.Y., who has made pore pressure measurements in the deep ocean.

T.F. LOMENICK, United States

I extend to you a final invitation for any additional comments.

F. GERA, NEA

It is true that there are not many in situ experiments at shallow depth being planned. The table shows four in situ experiments

IN SITU HEATING EXPERIMENTS

	Boom clay	Trisaia	Eleana	Conasauga	Seabed ISHTE
Heater					
- Power	3.6 kW (max)	1.2 kW	3.5 kW	7.0 kW (30 days) 4.5 kW (243 days)	400 W
- Length	1.5 m	2.0 m	3.0 m	3.0 m	0.45 m
- Diameter	0.3 m	0.2 m	0.3 m	0.31 m	0.08 m
- Material	S.S. 304	S.S.	S.S. 304	S.S. 304	Inconel 600
Duration	start ~1-6-79 (1 year minimum)	7 months	251 days	243 days	365 days
Depth	~ 6.5 m (median)	25 m	22.5 m	15 m	6000 m (water) 1.2 m (sediment)
Measurements	T° (continuously at 28 points) sampling + various lab. measurements	T° (continuously)	T°, permeability, gas permeability, displacement, stress	T°, permeability, displacement	T°, conductivity, pore pressure, sediment samples, photos
Geology	Tertiary plastic clay	Marly-clay of Pliocene-Pleistocene age	Argillite	Shale (siltstone, limestone)	Illitic clay
Laboratory	CEN/SCK B-2400 Mol Belgium	CNEN-CSN Casaccia Lab. Rif. Rad. C.P. 2400 00100 Rome Italy	Sandia Lab., Albuquerque, N.M. USA	Sandia Lab., Albuquerque, N.M. USA	Sandia Lab., Albuquerque, N.M. USA
Contact	R.H. Heremans	E. Tassoni	J.L. Krumhansl A.R. Lappin	L.D. Tyler J.L. Krumhansl	D.F. McVey

on land, that are now practically completed, and one in the seabed presently being planned. At the two extremes there are small-scale laboratory tests and in situ deep tests ; on the other hand, once the problem of measuring pore fluid pressure has been solved it might be worthwhile to consider a new generation of in situ experiments at shallow depth to see if the instrumentation works and what kind of information we get about the behaviour of pore fluids. Finally, I would like to ask Mr. McVey if he could tell us a little more about the heating experiment in deep marine clays.

D.F. McVEY, United States

It will take place in the early 1980's, approximately 800 miles north of Hawaii at a depth of about 6000 metres. We will use a three legged platform that is about twice as tall as a man. We will emplace a 400 watts heat source, that derives its power from four Pu-238 sources that were developed for space applications and have been approved for deep ocean use. The heater, 45 cm long and 8 cm in diameter, will be inserted directly into the sediments. The insertion mechanism, which will remain attached to the heater, is designed to minimize the heat flow from the heater back to the ocean floor. The surface of the insertion rod is treated to ensure a good contact with the clay sediments and a reasonably good sealing of the heated zone. Instrumentation will include about 80 thermal sensors arrayed in three plains spaced at 120° around the heater. The distance of the sensors from the heater will range between 2 and 200 cm so that isotherms will be accurately defined. In addition we will take cores both before and after the test, that is anticipated to last about one year. At the end of the test the heater will be drawn back into the platform, which will then be recovered from the bottom of the ocean. Six cores will be taken in the heated zone, in a radius of 75 cm. This should provide a fairly complete picture of what happened to the sediments during the experiment. In addition, before and after the test, we will measure the thermal conductivity of the sediments and, hopefully, the pore fluid pressures at three locations. The platform will carry cameras and a light system so that we can monitor the ocean floor to determine any additional biological activity or any visible effects of the test on the sediments. The mechanical design of the system has been described in a report published at the University of Washington.

J.B. LEWIS, United Kingdom

There is a different topic which should be mentioned ; that is the possible use of argillaceous deposits for disposal of medium- and low-level wastes. Some of the work discussed here can be useful for understanding better how to make use of argillaceous rocks for disposal of wastes that do not generate significant amounts of heat. Problems related to hydrology, radionuclides migration and so on are likely to be similar in both cases.

T.F. LOMENICK, United States

A very good point. At Oak Ridge we have disposed of intermediate-level liquid waste through hydraulic fracturing in a shale formation for more than a decade. The same technique could be used for high-level waste, if it were diluted to the point where heat generation would be no longer a problem. It would be interesting to discuss in some detail alternative schemes for disposal, but unfortunately there is very little time left.

F. GERA, NEA

Basically the performance of the geologic barrier is the same. The main difference is the absence of heat ; since heat is the main problem with high-level waste, disposal of other types of waste should be much easier. We are now learning about the performance of the geologic barrier and the migration rates of critical radionuclides. Without the disturbing influence of heat it should be relatively easy to come up with satisfactory migration models and convincing safety assessments. Several waste emplacement concepts are possible, hydraulic fracturing is only one of them. It would be possible to use argillaceous formations much closer to the surface than presently envisaged for high-level waste. Particularly for wastes that do not contain large amounts of alpha emitters, for they would not need to be isolated for extremely long periods of time. In cases where argillaceous formations are exposed at the surface, it might be possible to gain access to the formation with a subhorizontal tunnel. I think argillaceous sediments have great potential for waste disposal particularly for wastes that produce little decay heat.

T.F. LOMENICK, United States

One more thing that should be mentioned is the borehole matrix concept, which would seem to have some applicability for disposal in argillaceous formations.

If there are no more comments, I thank you all for your participation and close the session and the workshop.

LIST OF PARTICIPANTS

LISTE DES PARTICIPANTS

BELGIUM - BELGIQUE

BONNE, A.A., Centre d'Etude de l'Energie Nucléaire, Boeretang 200, B-2400 Mol

HEREMANS, R.H., Centre d'Etude de l'Energie Nucléaire, Boeretang 200, B-2400 Mol

FINLAND - FINLANDE

LUMIAHO, K.,Geological Survey of Finland, Kivimiehentie 1, SF-02150 Espoo 15

FRANCE

FEUGA, B., Bureau de Recherches Géologiques et Minières, B.P. 6009, 45018 Orléans Cedex

RANCON, D., Institut de Protection et de Sûreté Nucléaire, Commissariat à l'Energie Atomique, DSN/SESTR, Centre d'Etudes Nucléaires de Cadarache, B.P. n° 1, 13115 Saint-Paul-lez-Durance

ROCHON, J., Bureau de Recherches Géologiques et Minières, B.P. 6009, 45018 Orléans Cedex

ITALY - ITALIE

BRONDI, A., Comitato Nazionale per l'Energia Nucleare, Laboratorio Rifiuti Radioattivi, CSN Casaccia, 00060 S. Maria di Galeria, Rome

TASSONI, E., Comitato Nazionale per l'Energia Nucleare, Laboratorio Rifiuti Radioattivi, CSN Casaccia, 00060 S. Maria di Galeria, Rome

SWEDEN - SUEDE

JACOBSSON, A., Division of Soil Mechanics, University of Luleå, S-951 87 Luleå

PUSCH, R., Division of Soil Mechanics, University of Luleå, S-951 87 Luleå

SWITZERLAND - SUISSE

NOLD, A.L., Nationale Genossenschaft für die Lagerung Radioaktiver Abfälle, NAGRA, Parkstrasse 23, CH-5401 Baden

SCHMID, G., Dr., Colombi Schmutz Dorthe AG, Ziegelrain 29, CH-5000 Aarau

UNITED KINGDOM - ROYAUME-UNI

CHAPMAN, N.A., Dr., Institute of Geological Sciences, Building 151, Harwell Laboratory, Harwell, Oxfordshire, OX11 ORQ

DIRMIKIS, D., Dr., Dames & Moore Advanced Technology Group, 123 Mortlake High Street, London SW14 8SN

DUNCAN, A.G., Dr., Department of the Environment, Room 427, Becket House, 1 Lambeth Palace Road, London SE1 7ER

LEWIS, J.B., Dr., Atomic Energy Research Establishment, Building 175, AERE Harwell, Oxfordshire OX11 ORA

MARTI, J., Dames & Moore Advanced Technology Group, 123 Mortlake High Street, London SW14 8SN

UNITED STATES - ETATS-UNIS

BRUNTON, G.D., Oak Ridge National Laboratory, P.O. Box X, Oak Ridge, Tennessee 37830

DOLE, L.R., c/o Gesellschaft für Strahlen- und Umweltforschung mbH München, Institut für Tieflagerung Wissenschaftliche Abteilung, Berlinerstr. 2, D-3392 Clausthal-Zellerfeld, Federal Republic of Germany

ERDAL, B.R., Los Alamos Scientific Laboratory, CNC-11, MS-514, Los Alamos, New Mexico 87545

GONZALES, S., Dr., Earth Resource Associates, Inc., 125 Cedar Creek Drive, Athens, Georgia 30605

KRUMHANSL, J.L., Sandia Laboratories, Geological Protects Division 4537, P.O. Box 5800, Albuquerque, New Mexico 87185

LAPPIN, A.R., Sandia Laboratories, Geological Projects Division 4537, P.O. Box 5800, Albuquerque, New Mexico 87185

LOMENICK, T.F., Oak Ridge National Laboratory, P.O. Box X, Oak Ridge, Tennessee 37830

McVEY, D.F., Sandia Laboratories, Fluid Mechanics and Heat Transfer, Division I - 5511, P.O. Box 5800, Albuquerque, New Mexico 87185

WILSON, C.R., Lawrence Berkeley Laboratory, University of California, c/o Stripa Gruv AB, 710 50 Stora, Sweden

COMMISSION OF THE EUROPEAN COMMUNITIES
COMMISSION DES COMMUNAUTES EUROPEENNES

AVOGADRO, A., Dr., Chemical Division, Joint Research Center, Ispra (Varèse), Italy

MASURE, P., DG XII - D1, 200 rue de la Loi, 1049 Bruxelles, Belgium

SECRETARIAT

GERA, F., Dr., Division of Radiation Protection and Waste Management, Nuclear Energy Agency, 38 boulevard Suchet, 75016 Paris, France

SOME NEW PUBLICATIONS OF NEA

QUELQUES NOUVELLES PUBLICATIONS DE L'AEN

ACTIVITY REPORTS

RAPPORTS D'ACTIVITE

Activity Reports of the OECD Nuclear Energy Agency (NEA)

- 6th Activity Report (1977)
- 7th Activity Report (1978)

Rapports d'activité de l'Agence de l'OCDE pour l'Energie Nucléaire (AEN)

- 6ème Rapport d'Activité (1977)
- 7ème Rapport d'Activité (1978)

Free on request - Gratuit sur demande

Annual Reports of the OECD HALDEN Reactor Project

- 18th Annual Report (1977)
- 19th Annual Report (1978)

Rapports annuels du Projet OCDE de réacteurs de HALDEN

- 18ème Rapport annuel (1977)
- 19ème Rapport annuel (1978)

Free on request - Gratuit sur demande

Twentieth Anniversary of the OECD Nuclear Energy Agency

- Proceedings on the NEA Symposium on International Co-operation in the Nuclear Field : Perspectives and Prospects

Vingtième Anniversaire de l'Agence de l'OCDE pour l'Energie Nucléaire

- Compte rendu du Symposium de l'AEN sur la coopération internationale dans le domaine nucléaire : bilan et perspectives

Free on request - Gratuit sur demande

NEA at a Glance

Coup d'oeil sur l'AEN

Free on request - Gratuit sur demande

SCIENTIFIC AND TECHNICAL PUBLICATIONS

PUBLICATIONS SCIENTIFIQUES ET TECHNIQUES

NUCLEAR FUEL CYCLE

LE CYCLE DU COMBUSTIBLE NUCLEAIRE

Reprocessing of Spent Nuclear Fuels in OECD Countries

Retraitement du combustible nucléaire dans les pays de l'OCDE

1977
£ 2.50, US$ 5.00, F 20.00

Nuclear Fuel Cycle Requirements and Supply Considerations, Through the Long-Term

Besoins liés au cycle du combustible nucléaire et considérations sur l'approvisionnement à long terme

1978
£ 4.30, US$ 8.75, F 35.00

World Uranium Potential - An International Evaluation

Potentiel mondial en uranium - Une évaluation internationale

1978
£ 7.80, US$ 16.00, F 64.00

Uranium - Resources, Production and Demand

Uranium - Ressources, Production et Demande

1979
£ 8.70, US$ 19.50, F. 78.00

RADIATION PROTECTION

RADIOPROTECTION

Radon Monitoring (Proceedings of the NEA Specialist Meeting, Paris)

Surveillance du radon (Compte rendu d'une réunion de spécialistes de l'AEN, Paris)

1978
£ 8.00, US$ 16.50, F 66.00

Iodine-129 (Proceedings of an NEA Specialist Meeting, Paris)

Iode-129 (Compte rendu d'une réunion de spécialistes de l'AEN, Paris)

1977
£ 3.40, US$ 7.00, F 28.00

Recommendations for Ionization Chamber Smoke Detectors in Implementation of Radiation Protection Standards

Recommandations relatives aux détecteurs de fumée à chambre d'ionisation en application des normes de radioprotection

1977
Free on request - Gratuit sur demande

Management, Stabilisation and Environmental Impact of Uranium Mill Tailings
(Proceedings of the Albuquerque Seminar, United States)

Gestion, stabilisation et incidence sur l'environnement des résidus de traitement de l'uranium
(Compte rendu du Séminaire d'Albuquerque, Etats-Unis)

1978
£ 9.80, US$ 20.00, F 80.00

Exposure to Radiation from the Natural Radioactivity in Building Materials
(Report by an NEA Group of Experts)

Exposition aux rayonnements due à la radioactivité naturelle des matériaux de construction
(Rapport établi par un Groupe d'experts de l'AEN)

1979
Free on request - Gratuit sur demande

Marine Radioecology
(Proceedings of the Tokyo Seminar)

Radioécologie marine
(Compte rendu du Colloque de Tokyo)

1980
in preparation - en préparation

RADIOACTIVE WASTE MANAGEMENT

GESTION DES DECHETS RADIOACTIFS

Bituminization of Low and Medium Level Radioactive Wastes
(Proceedings of the Antwerp Seminar)

Conditionnement dans le bitume des déchets radioactifs de faible et de moyenne activités
(Compte rendu du Séminaire d'Anvers)

1976
£ 4.70, US$ 10.00, F 42.00

Objectives, Concepts and Strategies for the Management of Radioactive Waste Arising from Nuclear Power Programmes
(Report by an NEA Group of Experts)

Objectifs, concepts et stratégies en matière de gestion des déchets radioactifs résultant des programmes nucléaires de puissance
(Rapport établi par un Groupe d'experts de l'AEN)

1977
£ 8.50, US$ 17.50, F 70.00

Treatment, Conditioning and Storage of Solid Alpha-Bearing Waste and Cladding Hulls
(Proceedings of the NEA/IAEA Technical Seminar, Paris)

Traitement, conditionnement et stockage des déchets solides alpha et des coques de dégainage
(Compte rendu du Séminaire technique AEN/AIEA, Paris)

1977
£ 7.30, US$ 15.00, F 60.00

Storage of Spent Fuel Elements
(Proceedings of the Madrid Seminar)

Stockage des éléments combustibles irradiés
Compte rendu du Séminaire de Madrid)

1978
£ 7.30, US$ 15.00, F 60.00

In Situ Heating Experiments in Geological Formations (Proceedings of the Ludvika Seminar, Sweden)

Expériences de dégagement de chaleur in situ dans les formations géologiques (Compte rendu du Séminaire de Ludvika, Suède)

1978
£ 8.00, US$ 16.50, F 66.00

Migration of Long-lived Radionuclides in the Geosphere (Proceedings of the Brussels Workshop)

Migration des radionucléides à vie longue dans la géosphère (Compte rendu de la réunion de travail de Bruxelles)

1979
£ 8.30, US$ 17.00, F 68.00

Low-Flow, Low-Permeability Measurements in Largely Impermeable Rocks (Proceedings of the Paris Workshop)

Mesures des faibles écoulements et des faibles perméabilités dans des roches relativement imperméables (Compte rendu de la réunion de travail de Paris)

1979
£ 7.80, US$ 16.00, F 64.00

On-Site Management of Power Reactor Wastes (Proceedings of the Zurich Symposium)

Gestion des déchets en provenance des réacteurs de puissance sur le site de la centrale (Compte rendu du Colloque de Zurich)

1979
£ 11.00, US$ 22.50, F 90.00

Recommended Operational Procedures for Sea Dumping of Radioactive Waste

Recommandations relatives aux procédures d'exécution des opérations d'immersion de déchets radioactifs en mer

1979
Free on request - Gratuit sur demande

Guidelines for Sea Dumping Packages of Radioactive Waste (Revised version)

Guide relatif aux conteneurs de déchets radioactifs destinés au rejet en mer (Version révisée)

1979
Free on request - Gratuit sur demande

Use of Argillaceous Materials for the Isolation of Radioactive Waste (Proceedings of the Paris Workshop)

Utilisation des matériaux argileux pour l'isolation des déchets radioactifs (Compte rendu de la réunion de travail de Paris)

1980
in preparation - en préparation

SAFETY

SURETE

Safety of Nuclear Ships
(Proceedings of the Hamburg Symposium)

Sûreté des navires nucléaires
(Compte rendu du Symposium de Hambourg)

1978
£ 17.00, US$ 35.00, F 140.00

Nuclear Aerosols in Reactor Safety
(A State-of-the-Art Report by a Group of Experts)

Les aérosols nucléaires dans la sûreté des réacteurs
(Rapport sur l'état des connaissances établi par un Groupe d'Experts)

1979
£ 8.30, US$ 18.75, F 75.00

Plate Inspection Programme
(Report from the Plate Inspection Steering Committee - PISC - on the Ultrasonic Examination of Three Test Plates)

Programme d'inspection des plaques
(Rapport du Comité de Direction de l'inspection des plaques - PISC - sur l'examen par ultrasons de trois plaques d'essai)

1980
£ 3.30, US$ 7.50, F 30.00

SCIENTIFIC INFORMATION

INFORMATION SCIENTIFIQUE

Neutron Physics and Nuclear Data for Reactors and other Applied Purposes
(Proceedings of the Harwell International Conference)

La physique neutronique et les données nucléaires pour les réacteurs et autres applications
(Compte rendu de la Conférence Internationale de Harwell)

1978
£ 26.80, US$ 55.00, F 220.00

LEGAL PUBLICATIONS	PUBLICATIONS JURIDIQUES
Convention on Third Party Liability in the Field of Nuclear Energy - Incorporating the provisions of the Additional Protocol of January 1964	Convention sur la responsabilité civile dans le domaine de l'énergie nucléaire - Texte incluant les dispositions du Protocole Additionnel de janvier 1964

1960

Free on request - Gratuit sur demande

Nuclear Legislation, Analytical Study : "Nuclear Third Party Liability" (revised version)	Législations nucléaires, étude analytique : "Responsabilité civile nucléaire" (version révisée)

1976

£ 6.00, US$ 12.50, F 50.00

Nuclear Law Bulletin (Annual Subscription - two issues and supplements)	Bulletin de Droit Nucléaire (Abonnement annuel - deux numéros et suppléments)

£ 5.60, US$ 12.50, F 50.00

Index of the first twenty issues of the Nuclear Law Bulletin	Index des vingt premiers numéros du Bulletin de Droit Nucléaire

Free on request - Gratuit sur demande

Licensing Systems and Inspection of Nuclear Installations in NEA Member Countries (two volumes)	Régime d'autorisation et d'inspection des installations nucléaires dans les pays de l'AEN (deux volumes)

Free on request - Gratuit sur demande

NEA Statute	Statuts de l'AEN

Free on request - Gratuit sur demande

OECD SALES AGENTS
DÉPOSITAIRES DES PUBLICATIONS DE L'OCDE

ARGENTINA – ARGENTINE
Carlos Hirsch S.R.L., Florida 165, 4° Piso (Galería Guemes)
1333 BUENOS-AIRES, Tel. 33-1787-2391 Y 30-7122

AUSTRALIA – AUSTRALIE
Australia & New Zealand Book Company Pty Ltd.,
23 Cross Street, (P.O.B. 459)
BROOKVALE NSW 2100 Tel. 938-2244

AUSTRIA – AUTRICHE
Gerold and Co., Graben 31, WIEN 1. Tel. 52.22.35

BELGIUM – BELGIQUE
LCLS
44 rue Otlet, B1070 BRUXELLES. Tel. 02-521 28 13

BRAZIL – BRÉSIL
Mestre Jou S.A., Rua Guaipá 518,
Caixa Postal 24090, 05089 SAO PAULO 10. Tel. 261-1920
Rua Senador Dantas 19 s/205-6, RIO DE JANEIRO GB.
Tel. 232-07. 32

CANADA
Renouf Publishing Company Limited,
2182 St. Catherine Street West,
MONTREAL, Quebec H3H 1M7 Tel. (514) 937-3519

DENMARK – DANEMARK
Munksgaards Boghandel,
Nørregade 6, 1165 KØBENHAVN K. Tel. (01) 12 85 70

FINLAND – FINLANDE
Akateeminen Kirjakauppa
Keskuskatu 1, 00100 HELSINKI 10. Tel. 65-11-22

FRANCE
Bureau des Publications de l'OCDE,
2 rue André-Pascal, 75775 PARIS CEDEX 16. Tel. (1) 524.81.67
Principal correspondant :
13602 AIX-EN-PROVENCE : Librairie de l'Université.
Tel. 26.18.08

GERMANY – ALLEMAGNE
OECD Publications and Information Centre
4 Simrockstrasse
5300 BONN Tel. 21 60 46

GREECE – GRÈCE
Librairie Kauffmann, 28 rue du Stade,
ATHÈNES 132. Tel. 322.21.60

HONG-KONG
Government Information Services,
Sales and Publications Office, Beaconsfield House, 1st floor,
Queen's Road, Central. Tel. 5-233191

ICELAND – ISLANDE
Snaebjörn Jónsson and Co., h.f.,
Hafnarstraeti 4 and 9, P.O.B. 1131, REYKJAVIK.
Tel. 13133/14281/11936

INDIA – INDE
Oxford Book and Stationery Co.:
NEW DELHI, Scindia House. Tel. 45896
CALCUTTA, 17 Park Street. Tel. 240832

ITALY – ITALIE
Libreria Commissionaria Sansoni:
Via Lamarmora 45, 50121 FIRENZE. Tel. 579751
Via Bartolini 29, 20155 MILANO. Tel. 365083
Sub-depositari:
Editrice e Libreria Herder,
Piazza Montecitorio 120, 00 186 ROMA. Tel. 674628
Libreria Hoepli, Via Hoepli 5, 20121 MILANO. Tel. 865446
Libreria Lattes, Via Garibaldi 3, 10122 TORINO. Tel. 519274
La diffusione delle edizioni OCSE è inoltre assicurata dalle migliori librerie nelle città più importanti.

JAPAN – JAPON
OECD Publications and Information Center
Akasaka Park Building, 2-3-4 Akasaka, Minato-ku,
TOKYO 107. Tel. 586-2016

KOREA - CORÉE
Pan Korea Book Corporation,
P.O.Box n° 101 Kwangwhamun, SÉOUL. Tel. 72-7369

LEBANON – LIBAN
Documenta Scientifica/Redico,
Edison Building, Bliss Street, P.O.Box 5641, BEIRUT.
Tel. 354429–344425

MALAYSIA – MALAISIE
University of Malaya Co-operative Bookshop Ltd.
P.O. Box 1127, Jalan Pantai Baru
KUALA LUMPUR Tel. 51425, 54058, 54361

THE NETHERLANDS – PAYS-BAS
Staatsuitgeverij
Verzendboekhandel
Chr. Plantijnstraat
'S-GRAVENHAGE Tel. nr. 070-789911
Voor bestellingen: Tel. 070-789208

NEW ZEALAND – NOUVELLE-ZÉLANDE
The Publications Manager,
Government Printing Office,
WELLINGTON: Mulgrave Street (Private Bag),
World Trade Centre, Cubacade, Cuba Street,
Rutherford House, Lambton Quay, Tel. 737-320
AUCKLAND: Rutland Street (P.O.Box 5344), Tel. 32.919
CHRISTCHURCH: 130 Oxford Tce (Private Bag), Tel. 50.331
HAMILTON: Barton Street (P.O.Box 857), Tel. 80.103
DUNEDIN: T & G Building, Princes Street (P.O.Box 1104),
Tel. 78.294

NORWAY – NORVÈGE
J.G. TANUM A/S
P.O. Box 1177 Sentrum
Karl Johansgate 43
OSLO 1 Tel (02) 80 12 60

PAKISTAN
Mirza Book Agency, 65 Shahrah Quaid-E-Azam, LAHORE 3.
Tel. 66839

PORTUGAL
Livraria Portugal, Rua do Carmo 70-74,
1117 LISBOA CODEX.
Tel. 360582/3

SPAIN – ESPAGNE
Mundi-Prensa Libros, S.A.
Castelló 37, Apartado 1223, MADRID-1. Tel. 275.46.55
Libreria Bastinos, Pelayo, 52, BARCELONA 1. Tel. 222.06.00

SWEDEN – SUÈDE
AB CE Fritzes Kungl Hovbokhandel,
Box 16 356, S 103 27 STH, Regeringsgatan 12,
DS STOCKHOLM. Tel. 08/23 89 00

SWITZERLAND – SUISSE
Librairie Payot, 6 rue Grenus, 1211 GENÈVE 11. Tel. 022-31.89.50

TAIWAN – FORMOSE
National Book Company,
84-5 Sing Sung Rd., Sec. 3, TAIPEI 107. Tel. 321.0698

THAILAND – THAILANDE
Suksit Siam Co., Ltd.
1715 Rama IV Rd.
Samyan, Bangkok 5
Tel. 2511630

UNITED KINGDOM – ROYAUME-UNI
H.M. Stationery Office, P.O.B. 569,
LONDON SE1 9 NH. Tel. 01-928-6977, Ext. 410 or
49 High Holborn, LONDON WC1V 6 HB (personal callers)
Branches at: EDINBURGH, BIRMINGHAM, BRISTOL,
MANCHESTER, CARDIFF, BELFAST.

UNITED STATES OF AMERICA – ÉTATS-UNIS
OECD Publications and Information Center, Suite 1207,
1750 Pennsylvania Ave., N.W. WASHINGTON, D.C. 20006.
Tel. (202)724-1857

VENEZUELA
Libreria del Este, Avda. F. Miranda 52, Edificio Galipán,
CARACAS 106. Tel. 32 23 01/33 26 04/33 24 73

YUGOSLAVIA – YOUGOSLAVIE
Jugoslovenska Knjiga, Terazije 27, P.O.B. 36, BEOGRAD.
Tel. 621-992

Les commandes provenant de pays où l'OCDE n'a pas encore désigné de dépositaire peuvent être adressées à :
OCDE, Bureau des Publications, 2 rue André-Pascal, 75775 PARIS CEDEX 16.
Orders and inquiries from countries where sales agents have not yet been appointed may be sent to:
OECD, Publications Office, 2 rue André-Pascal, 75775 PARIS CEDEX 16.

PUBLICATIONS DE L'OCDE
2, rue André-Pascal
75775 PARIS CEDEX 16
N° 41 477 1980
(1700 SH 66 80 03 3) ISBN 92-64-02050-0

●

IMPRIMÉ EN FRANCE